Introductory Statistics for Biology Students

Introductory Statistics for Biology Students

T.A. Watt
Department of Agriculture, Horticulture and the Environment,
Wye College, University of London, Wye, UK

CHAPMAN & HALL
London · Glasgow · New York · Tokyo · Melbourne · Madras

Published by Chapman & Hall, 2–6 Boundary Row, London SE1 8HN

Chapman & Hall, 2–6 Boundary Row, London SE1 8HN, UK

Blackie Academic & Professional, Wester Cleddens Road, Bishopbriggs, Glasgow G64 2NZ, UK

Chapman & Hall Inc., 29 West 35th Street, New York NY10001, USA

Chapman & Hall Japan, Thomson Publishing Japan, Hirakawacho Nemoto Building, 6F, 1-7-11 Hirakawa-cho, Chiyoda-ku, Tokyo 102, Japan

Chapman & Hall Australia, Thomas Nelson Australia, 102 Dodds Street, South Melbourne, Victoria 3205, Australia

Chapman & Hall India, R. Seshadri, 32 Second Main Road, CIT East, Madras 600 035, India

First edition 1993

Typeset in 10/12pt Times by Falcon Graphic Art Limited, Surrey

Printed in Great Britain by Page Bros., Norwich

ISBN 0 412 47150 7

A catalogue record for this book is available from the British Library

Library of Congress Cataloging-in-Publication data available

∞ Printed on permanent acid-free text paper, manufactured in accordance with the proposed ANSI/NISO Z 39.48-199X and ANSI Z 39.48-1984

To Colyear Dawkins

Contents

Preface

Students of biological subjects in further and higher education often lack confidence in their numerical abilities. At the beginning of their course mature students commonly describe their experience in dealing with numbers as 'rusty', while those students coming straight from school may have few science A-levels and often have only a modest pass in GCSE mathematics. Many individuals from both groups express their surprise at having to study statistics as part of their course in biological sciences.

This book aims to calm their fears; to show why biologists need statistics and to provide a painless way into the subject. MINITAB *Statistical Software* removes drudgery, allows easy data checking and exploratory analysis and enables complex analyses to be undertaken with ease. [MINITAB is a registered trademark of Minitab Inc. whose cooperation I gratefully acknowledge.*]

My own interest in statistics was inspired by the late Colyear Dawkins whose enthusiasm for communicating the principles of experimental design and analysis was infectious. I also value discussions with many colleagues over the years, including: Alan Clewer, John Palmer, Jean Power and Howard Wright. I am grateful to Brian Carpenter, Keith Kirby, Mark Macnair, Dominic Recaldin and Eunice Simmons for their constructive comments on draft material.

*Minitab Inc. is based at 3081 Enterprise Drive, State College, PA 16801-3008, USA. Telephone: 0101-814-238-3280, fax: 0101-814-238-4383.

Note to students

Statistics is an essential tool for all life scientists. Unfortunately, a first-year statistics course at college or University is often seen as an inconvenient hurdle which must be jumped. Mathematical equations can appear daunting, so why bother? Because statistics really will be useful and important to you as this book aims to show. Furthermore I aim to cover the important ideas of statistics without using forbidding mathematical equations.

How long is a worm?

1

I would not enter on my list of friends . . . the man who needlessly sets foot upon a worm

William Cowper

1.1 INTRODUCTION

School experiments in physics and chemistry often have known answers. If you don't record a value of 9.8 metres per second per second for 'the acceleration with which an object falls to the earth' then you know it must be because there was something wrong with your equipment or with how you used it. Similarly, the molar mass of calcium carbonate is 100.09, so any other value would be wrong. The idea that there is a single clear-cut answer to a question isn't relevant in biology. 'How heavy is a hedgehog?' or 'what is the length of an earthworm?' do not have just one answer. However we need to know the answers because, for example, the weight of a hedgehog in autumn largely determines whether or not it will survive its over-winter hibernation. The aim of this chapter is to show how to obtain a useful answer to such questions.

We will simplify life by concentrating on just those earthworms of one species living in one particular field. Since earthworms are both male and female at the same time we don't need to specify which sex we wish to measure. Even so, individuals occur with widely differing lengths. Why is this? Like all animals, earthworms vary in their genetic makeup – some inherit a tendency to be short and fat and others to be long and thin. Earthworms can live for a long time and young worms are likely to be shorter than old ones.

Those which live in the moister part of the field at the bottom of the slope might be more active and have a better food supply so they will grow more quickly and may tend to be longer than those in a less favourable part of the field. Meanwhile perhaps those living near the footpath along one side of the field tend to be shorter because they are infested with a parasite or because they have recently escaped from a tussle with a bird. How then should we measure and describe the length of worms in this field?

1.2 SAMPLING A POPULATION

We would like to obtain information about *all* the worms in the field because they form the population in which we are interested, but it is impossible to measure them all. Therefore we measure a few – our sample – and generalize the results to the whole population. This is only a valid solution provided that the worms in our sample are representative of the whole population.

1.2.1 Measuring worms

Let's imagine that we have collected ten worms and propose to measure them, using a ruler, before returning them to their habitat. First, how do we make one keep straight and still? We will need to coax it into a straight line with one end at 0 mm and read off the measurement. We must avoid stretching it or allowing it to contract too much. This is difficult and will need practice. No doubt slime or soil will add to the problem of deciding the correct reading. I suspect that, although we would like to say that 'this particular worm is 83 mm long' and mean it, we will realize that this is unlikely to be a very accurate measurement.

Before you can even start to worry about analysing your results, therefore, you should think carefully about the errors and uncertainties involved in actually making the measurements. Would you be quite so careful in measuring worms, for example, on a dripping wet day as when the sun was shining?

We can but do our best to standardize the process and, in view of the various uncertainties, decide to measure earthworms only to the nearest 5 mm, i.e. 0.5 cm. Here are the results in cm:

11.5, 10.0, 9.5, 8.0. 12.5, 13.5, 9.5, 10.5, 9.0, 6.0

If we arrange our measurements in order we see at once the enormous variation in length:

6.0, 8.0, 9.0, 9.5, 9.5, 10.0, 10.5, 11.5, 12.5, 13.5

Some of the worms are twice as long as others. It is sensible to check our set of observations at this stage. For example, if one of them was 70 cm, do we remember measuring this giant (worms this size do exist in Australia), or is it more likely that we have misread 10 cm as 70 cm?

So how long is a worm? We could work out the **average** or **mean** length of our ten worms and say that it was the length of a 'typical' worm in our sample. If we add up the lengths and divide the sum by ten we get 10 cm. Is this, however, the length of a 'typical' worm in the field?

Let's imagine that our sample of ten worms had shown that each worm measured 10 cm. This is very consistent (not to mention very suspicious –

we should check that someone is not making up the results). The mean length is 10 cm and it would seem, in common-sense terms, very likely that the mean length of all the worms in the field (the population) is very close to this value. If we had sampled 20 worms and they were all correctly measured as 10 cm this would be amazingly strong evidence that the **population mean** is very close to 10 cm. The more worms we measure (increasing **replication**), the more information we have and so the greater our confidence that the **estimate** of the mean that we have obtained from the sample is close to the real population mean which we want to know but cannot measure directly.

However, the ten worms in our sample were **not** all the same length. Here they are again:

6.0, 8.0, 9.0, 9.5, 9.5, 10.0, 10.5, 11.5, 12.5, 13.5

The mean length is:

$$(6.0 + 8.0 + \ldots + 12.5 + 13.5)/10 = 100/10 = 10 \text{ cm}$$

With this amount of variation within the ten values, how confident can we be that the mean length of **all the worms in the field** is 10 cm? To answer this question we need a way of expressing the **variability** of the sample and using this as an estimate of the variability of the population, and for that you need 'statistics'.

1.2.2 How reliable is our sample estimate?

Statistical methods allow you to obtain an estimate of some characteristic (for example, the mean length of the worms) of a large population (the whole field) from only a small sample (ten worms). The **population's** characteristic is called a **parameter**, while the estimate of it, obtained from the **sample**, is called a **statistic**. So, if the mean length of all the worms in the field is actually 9.8 cm what we actually get from our measurements is an estimate of this parameter. First we took a sample of ten worms and calculated a sample mean of 10 cm. This is the sample statistic. We could of course take another sample of ten worms, in which case we would probably obtain a different estimate, perhaps 9.7 cm and so we might go on.

However, we usually only take one sample and so have only one sample statistic so it is important to be able to express the reliability of our measurements in estimating the parameter (the real mean length of all the worms). The following sections outline how this is done.

1.3 THE NORMAL DISTRIBUTION

If we summarize the information about the lengths of our ten worms in a graph, we see that there is a tendency for a large number of worms to be of medium length and fewer to be either very big or very small. Look at Fig. 1.1a, where the **frequency** of occurrence (or actual number) of worms in a range of size classes is shown. If we measure more worms, the number in the sample increases, and so the shape of the graph becomes less ragged (Fig. 1.1b). For the whole population of worms in the field (assuming that the field does not contain any localized abnormality) it is likely to follow a smooth 'bell-shaped' curve (Fig. 1.2), called the **Normal distribution**. (Normal is a technical word here – it doesn't just mean 'ordinary'.)

The spread of the curve depends on the natural variability in the population, being greater if the variability is large, i.e. there are relatively more values a long way from the centre line. This is a common shape of distribution for measurements like length and weight. However, other types of measurement, other populations, do not follow the normal distribution. For example, counts of the number of butterflies per thistle might produce mainly zeros (most thistles lack butterflies), a few ones and twos and the occasional larger number where the thistle is in a sunny spot and attracts many butterflies. Such a distribution would be **skewed** to the right (Fig. 1.3).

Instead of having a scale of frequency of worms up the side of our graph, we could have one of **probability** (how likely it is that a worm's length will fall into a particular length class). This probability scale would go from 0 to 1. If all worms in the field were 10 cm long, the probability that a worm is in the size class 9.5–10.4 would be 1. This represents 100% certainty. However, in reality the worms vary in length so the different size classes have different probabilities, with the highest being for the central size class

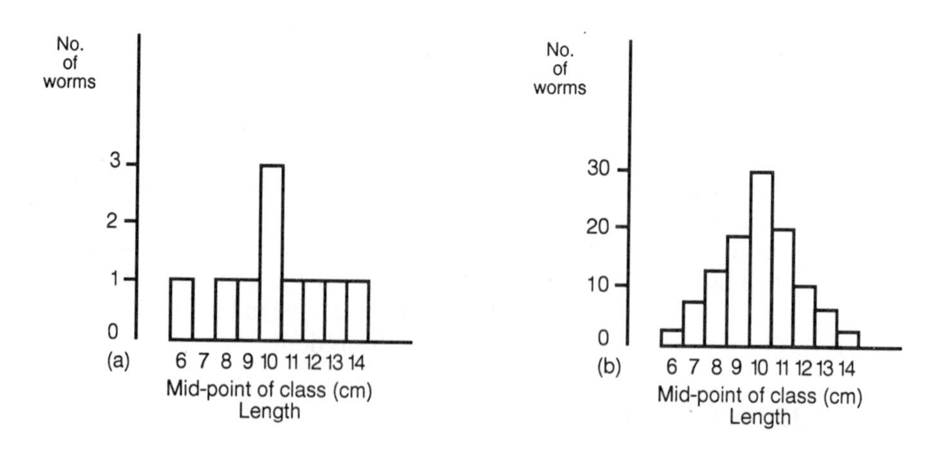

Figure 1.1 The number of worms in each size class. (a) Few worms (10 in our sample); (b) many worms.

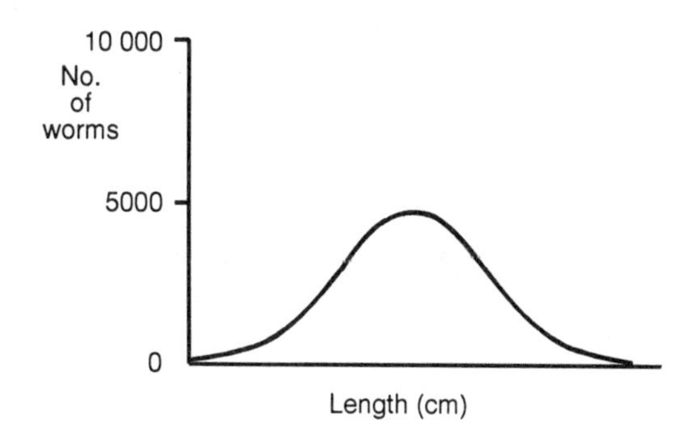

Figure 1.2 The Normal distribution.

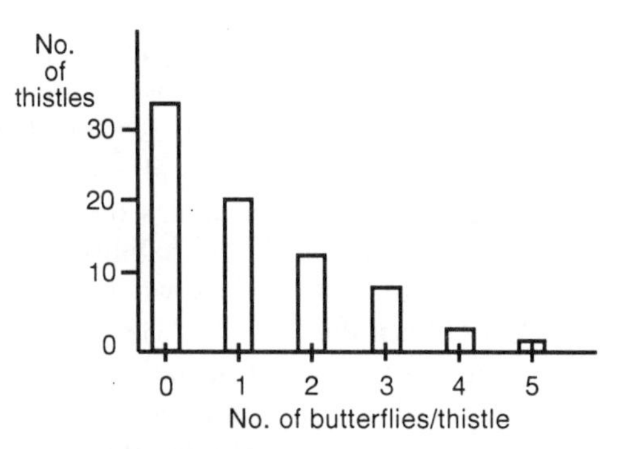

Figure 1.3 A positively-skewed distribution.

as in the Normal distribution curve in Fig. 1.2. The convenient aspect of this curve is that mathematicians have worked out how to describe it as an equation. The equation takes into account the mean (the central, highest point) and also its spread through an estimate of variability of the values from one individual to another – the **standard deviation**. We will see how to calculate the standard deviation from a sample of observations a few pages further on. For the moment it is enough to know that the larger the variability of a population the larger is the standard deviation.

If the equation of a curve is known, a mathematical technique (integration) can be used to work out the area underneath the curve, which represents the whole population. In addition, integration can be used to find out the area between any two length values. The ratio of such an area to the total area under the curve (representing the whole population) gives you the probability that you will encounter worms of a particular length. For example, if we wanted to know the probability of a worm being between 6 cm and 7 cm long we would find that it is 0.08 (or only 8% of the population, Fig. 1.4).

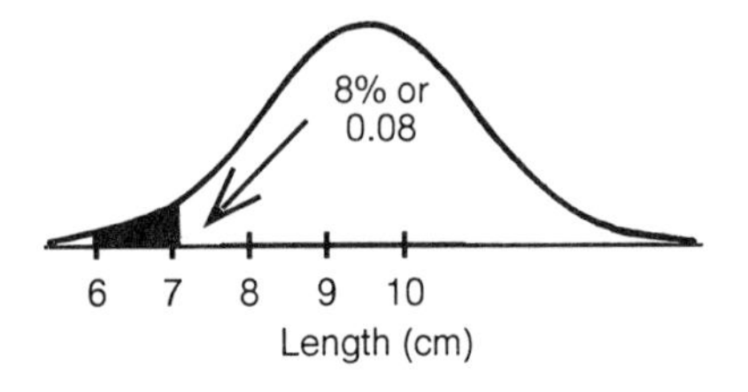

Figure 1.4 The probability of a worm being between 6 and 7 cm in length.

Fortunately you do not need to know anything about either the equation of the curve or about integration to be able to answer such questions. The information needed is already available in tables (Statistical tables) and most statistically orientated computer programs already incorporate them, so that we don't even need to consult the tables themselves very often. However, it is useful to remember that the probability of a worm's length being in the range between the two values where the curve changes direction (points of inflection) is about 0.68 (68%) (Fig. 1.5). These values are the mean plus or minus one standard deviation. For example, if the mean is 10 cm and the standard deviation was calculated to be 0.5 cm we would know that 68% of worms would have lengths between 9.5 cm and 10.5 cm. If the population was more variable and had a standard deviation of 2 cm we would expect 68% of worms to have lengths between 8 cm and 12 cm (Fig. 1.6).

How in practice therefore do we summarize the variability of a population in a way that will tell us something about the spread of the curve, so that we can work out the probability of finding worms of different lengths?

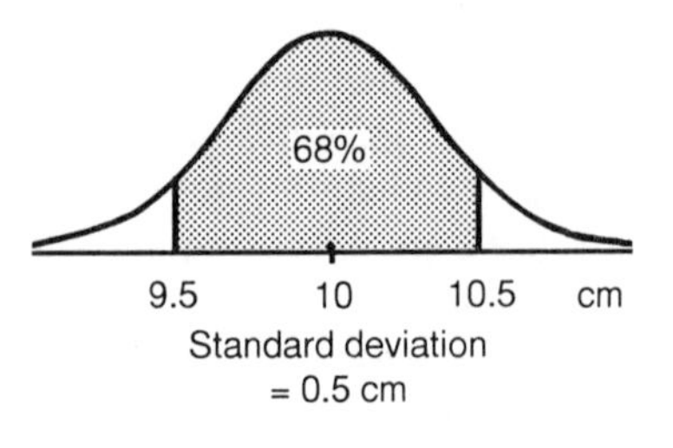

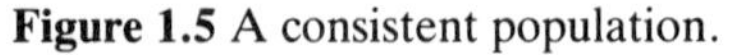
Figure 1.5 A consistent population.

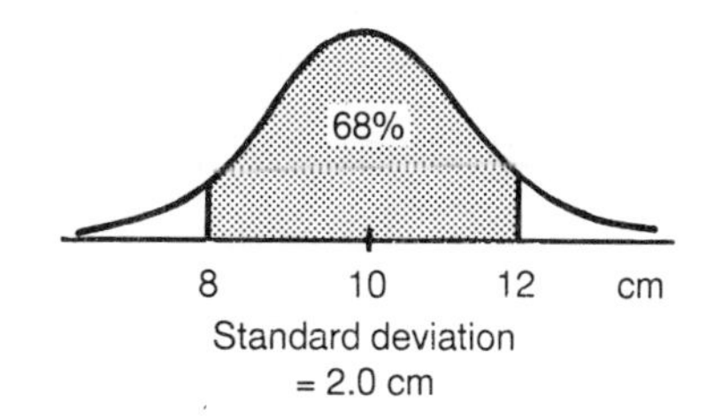

Figure 1.6 A variable population.

1.4 EXPRESSING VARIABILITY

1.4.1 First step – the sum of the differences

A first attempt to describe the variability of our initial sample of ten worms is to note how far away each value is from the mean and then to add up these distances or differences. It doesn't matter whether the values are greater or less than the mean, only how far away they are, so all the differences are treated as positive values in this exercise. The further away the values are from the mean, the greater their sum.

a Observation (cm)	b Distance from the mean value (cm)	c Difference (positive) (cm)
6.0	−4.0	4.0
8.0	−2.0	2.0
9.0	−1.0	1.0
9.5	−0.5	0.5
9.5	−0.5	0.5
10.0	0.0	0.0
10.5	0.5	0.5
11.5	1.5	1.5
12.5	2.5	2.5
13.5	3.5	3.5
Sum	0.0	16.0

Adding up the values in column b always gives a sum of zero, because by definition the mean is the value at which the sum of the distances below it equal those above it. However, when the differences are all treated as positive (often called **absolute** values) their sum is, in this instance, 16.

If five of the worms in our sample were each 9 cm long and the other five were 11 cm long, this method would give a mean of 10 cm. Intuitively we might consider such a set of observations less variable. What would be the sum of differences? (Fig. 1.7). So, the less the variability in the observa-

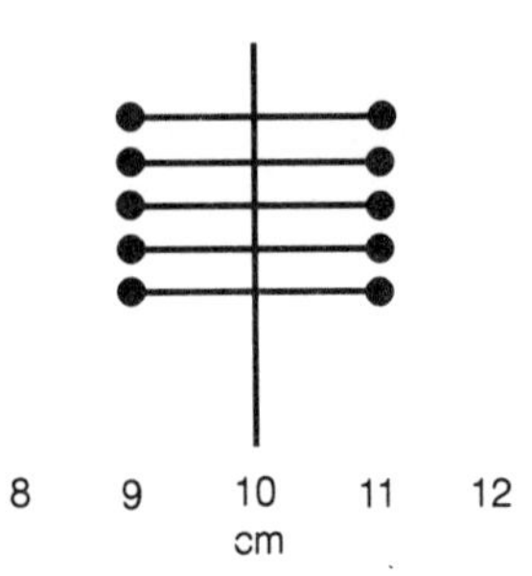

Figure 1.7 Ten differences of 1 cm.

tions, the smaller the sum of the differences. This seems a promising way of summarizing variability but we can improve upon it.

1.4.2 Second step – the sum of the squared differences

By our sum of the differences method, a sample containing just two worms, one of 6 cm and the other of 14 cm in length, is no more variable than one which has four worms of 9 cm and four worms of 11 cm. Both have a sum of differences of 8 (Fig. 1.8). But, from looking at the values, the first sample would seem to be more variable than the second.

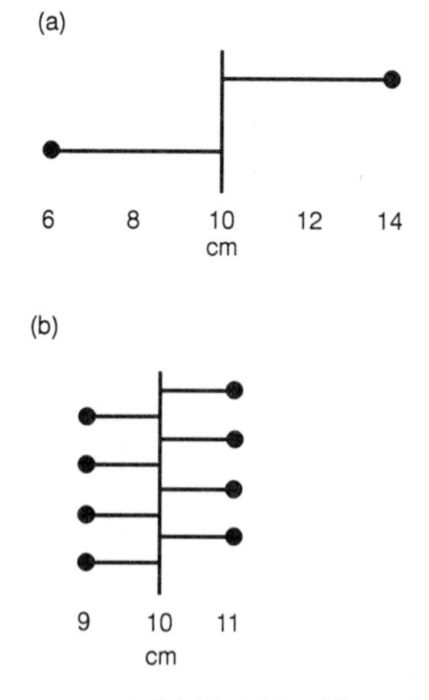

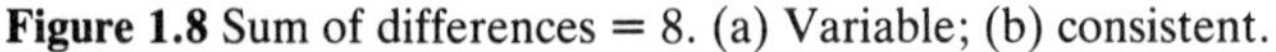

Figure 1.8 Sum of differences = 8. (a) Variable; (b) consistent.

In the second sample we have many observations which all fall only a small distance from the mean, which is evidence of consistency, whereas in the first sample the results suggest inconsistency; there are only a few values and these fall far away from the mean. A neat way of taking this consistency into account is to square each difference (to multiply it by itself) before adding them up. This means that extreme values will contribute more to the total sum.

Sample 1 measured values (cm)	Difference (all positive)	Difference squared
6	4	16
14	4	16
Sum	8	32
Sample 2 measured values (cm)		
9	1	1
9	1	1
9	1	1
9	1	1
11	1	1
11	1	1
11	1	1
11	1	1
Sum	8	8

The sum of the squared differences between the observations and their mean is 8 for the consistent sample (sample 2) but 32 for the inconsistent one (sample 1) so this revised method gives a much better picture of the variability present than just the sum of the differences used earlier. The phrase 'the sum of the squared differences between the observations and their mean' is usually abbreviated to 'the sum-of-squares'. However, this abbreviation sometimes leads to it being miscalculated. The 'sum-of-squares' does **not** mean the sum of the squares of each observation as the following example should make clear.

The sum-of-squares for a sample of three worms measuring 2, 3, and 4 cm respectively is not $(2\times2) + (3\times3) + (4\times4) = 29$. Instead, the mean of the values 2, 3 and 4 is 3, the sum-of-squares is:

$$(2-3)^2 + (3-3)^2 + (3-4)^2 = 1^2 + 0^2 + 1^2 = 2.$$

1.4.3 Third step – the variance

So far so good, but we have not allowed for the number of worms which we have in our sample in our calculation. In general we can refer to this as the number of **observations** (each worm measured counts as an observation). The more worms we measure the greater will be the 'sum-of-squares' simply because we are adding up more squared differences. To take account of this fact and so to provide a standard estimate of variability (called the variance), we divide the sum-of-squares by the number of observations minus one. In this case, we divide the sum-of-squares by 9 if we have ten worms or by 7 if we have eight worms. Why do we divide by one less than the number of observations rather than by the number itself?

If we have a sample of only one worm and it is 9 cm long, we have an estimate of the population mean (9 cm), but we have no information about how variable or unreliable this estimate is. As soon as we select a second worm (of length 10 cm) we have two values (9 and 10 cm). We can revise our estimate of the population mean to $(9 + 10)/2 = 9.5$ cm and we have one piece of information about its variability. This is the difference between the lengths of the two worms. With three worms we can revise the estimate of the mean again and we have two **independent** pieces of information about the variability:

Worm lengths: 9.0, 10.0, 11.0 New mean = 10.0

The two independent pieces of information about variability are (10.0–9.0) and (11.0–10.0). The third difference (11.0–9.0) is not independent because both values already feature as part of the other differences.

It is common to refer generally to the number of observations in a sample as n. So, in our main example, where $n = 10$, we divide the sum-of-squares of our sample by n–1 (which is 9) to obtain a **variance**. The number n–1 is referred to as the **degrees of freedom** because it refers to the number of independent items of information about variability that we can derive from the sample.

To reinforce the above justification for using (n–1) to calculate the variance statisticians have shown (using many lines of algebra) that this standard estimate of variability, the variance, calculated from a sample by dividing the sum-of-squares by n–1 is **unbiased**. This means that if we took many samples and calculated for each a variance value in this way, then their average would be close to the real population variance; it would be neither consistently too big nor consistently too small.

1.4.4 Fourth step – the standard deviation

We have now obtained variances (measures of variability within a sample) which can be compared with one another, but they are in units which are different from those used in the initial measurement. If you go back to section 1.4.2 you will see that we took the differences in length (cm) and then squared them. Thus in the case of our example the units of variance are in cm^2. We wouldn't naturally think of the variability of the length of a worm in such units. So we take the square root of the variance to return to cm. The result is a statistic called the **standard deviation**. This was referred to earlier in connection with the Normal distribution curve.

We can now work out the standard deviation from our original data set:

Observation	Difference (all positive)	Difference squared
6.0	4.0	16.0
8.0	2.0	4.0
9.0	1.0	1.0
9.5	0.5	0.25
9.5	0.5	0.25
10.0	0.0	0.0
10.5	0.5	0.25
11.5	1.5	2.25
12.5	2.5	6.25
13.5	3.5	12.25
Sum	16.0	42.5

The sum-of-squares is 42.5, so the variance is $42.5 \div 9 = 4.722$ and the standard deviation, which is the square root of this, is 2.173. This value helps us to judge the extent to which worm length varies from one individual to another, but its usefulness becomes clearer when we put this piece of information together with our knowledge of the Normal distribution (Fig. 1.5). This told us that 68% of the worms in the field will have lengths between:

the mean ± the standard deviation.

In our sample of ten worms that is between:

10 cm – 2.173 cm and 10 cm + 2.173 cm

which works out to be between

7.827 cm and 12.173 cm

This sounds amazingly precise – it implies that we can measure the length of a worm to the nearest 0.01 of a millimetre. Since we only measured our worms to the nearest 0.5 cm (5 mm) it is better to express this result as:

68% of the worms in the field have lengths between 7.8 cm and 12.2 cm.

If the worms we measured still had a mean length of 10 cm but had been much less variable, for example, with lengths mainly in the range from 9 to 11 cm, the standard deviation would be much less. Try working it out for these values:

8.5, 9.0, 9.5, 9.5, 10.0, 10.0, 10.5, 10.5, 11.0, 11.5
(you should get 0.913)

1.4.5 Fifth step – the sampling distribution of the mean

Measuring ten worms gives us an estimate of their mean length and of their variability but we have already implied that if we took a second sample of worms the estimate of the mean would be slightly different. Thus in this section we look at how the mean varies from one sample to another. This is referred to as a sampling distribution.

To illustrate what this means imagine that the population that we wish to sample consists of just six worms and that we are going to estimate the mean length of a worm by sampling just two of them at a time (the worms being returned to the population after each sampling). We could find the following possible outcomes:

Population (mean = 10)
Lengths 8 9 10 10 11 12

Sample	A	B	C	D	E	F	G	H	I	J	K	L	M	N	O
Worm 1	8	8	8	8	8	9	9	9	9	10	10	10	10	10	11
Worm 2	9	10	10	11	12	10	10	11	12	10	11	12	11	12	12

For convenience, where the two worms in a sample differ in length, the shorter has been called worm 1. There are 15 possible samples (A to O), because that is the maximum number of different combinations of pairs of worms that you can get from six worms. Because two of the worms share the same length (10 cm) some samples will give the same mean (see samples B and C, for example).

We can see that, by chance, our sample might have provided us with an estimate of the mean which was rather extreme: sample A would give 8.5, while sample O would give 11.5, compared with a population mean of 10.0. The means of the 15 possible samples can be summarized as follows:

Sample numbers	Number of samples	Sample mean	Number of samples × sample mean
A	1	8.5	8.5
B, C	2	9.0	18.0
D, F, G	3	9.5	28.5
E, H, J	3	10.0	30.0
I, K, M	3	10.5	31.5
L, N	2	11.0	22.0
O	1	11.5	1.5
	Total = 15		Total = 150
			Mean = 10.0
			= 150/15

The second column in this table shows that if we take a sample of two worms we have a 1 in 15 chance of selecting sample A and so getting an estimated mean of 8.5. However, we have a 3 in 15 (or 1 in 5) chance of having a sample with a mean of 9.5 (samples D, F or G) or of 10.0 (E, H or

J) or of 10.5 (I, K or M). This is the sampling distribution of the sample mean. It shows us that we are more likely to obtain a sample with a mean close to the true population mean than we are to obtain one with a mean far away from the population mean.

Also, the distribution is symmetrical (Fig. 1.9) and follows the Normal curve so that if we take a series of such samples we will obtain an unbiased estimate of the population mean. The mean of all 15 samples is of course the population mean (10.0, bottom of right-hand column) because all worms in the population have been sampled equally, with each worm occurring once with each of the other five worms in the population.

1.4.6 Sixth step – the standard error of the mean

In reality, we will only have one sample mean and we will not know the true population mean because that is one of the things we usually want to estimate. Therefore we need a way of expressing the variability from one sample mean to another. For example it is very common in **experiments** (Chapter 4) to want to compare say the mean length of ten worms fed on one type of food with the mean length of ten worms fed on a different type of food.

In earlier steps we showed how the standard deviation could be used to describe the variability from one worm to another. The standard deviation is obtained by taking the square root of the variance. It describes the dispersion of our observations. To estimate the variability of a mean a slightly different statistic is used – the **standard error**. In contrast to the standard deviation the standard error describes our uncertainty about the mean length of a worm.

This uncertainty is caused by the fact that if we took several samples, as we have illustrated above, they may each provide a different estimate of the mean. To obtain the standard error (of the mean) we divide the

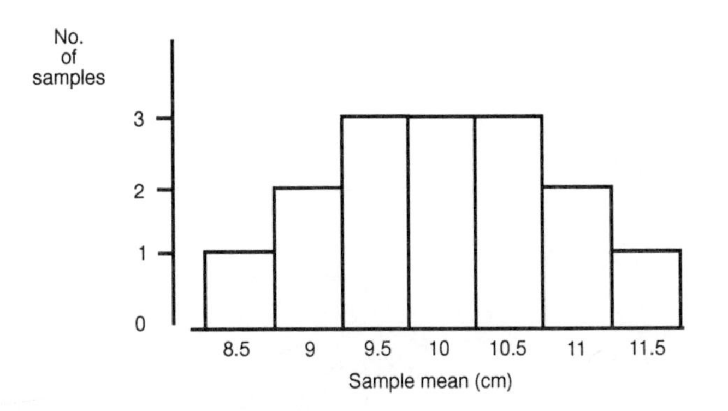

Figure 1.9 Means from the 15 possible samples each of two worms.

variance by the sample size before taking the square root. This can be expressed in three equivalent ways but mathematically they are all the same and give the same result:

$$\sqrt{\frac{\text{variance}}{\text{sample size}}} = \frac{\sqrt{\text{variance}}}{\sqrt{\text{sample size}}} = \frac{\text{standard deviation}}{\sqrt{\text{sample size}}}$$

The standard error gets smaller as the sample size increases because the standard deviation is divided by the square root of the sample size. So, if our sample size was 16 the standard error of the mean would be the standard deviation divided by four. If we took a sample of 25 we would divide the standard deviation by five and so on. This is because the lengths of occasional tiny or huge worms have much less influence on the mean when many worms are measured than if only a few are measured. Thus the mean of a larger sample is more likely to be closer to the population mean.

Back in step 4 the standard deviation of the sample of ten worms was found to be 2.173 cm. The standard error of the mean will be less because we divide the standard deviation by the square root of 10 (which is 3.162).

2.173/3.162 = 0.69. (we should perhaps round this to 0.7 cm)

Because the distribution of sample means follows the Normal curve then there is a 68% chance that the range between the estimated mean plus one standard error and the estimated mean minus one standard error will contain the population mean. So if the mean is 10.0 cm and the standard error of the mean is 0.7 cm there is a 68% chance that the range between 10 cm **plus** 0.7 cm and 10 cm **minus** 0.7 cm, i.e. from 9.3 cm to 10.7 cm contains the **population mean**.

Confidence intervals and computers

2

2.1 THE IMPORTANCE OF CONFIDENCE INTERVALS

Any estimate that we make of the mean value of a population mean should be accompanied by an estimate of its variability – the **standard error**. As we have seen in Chapter 1 this can be used to say within what range of values there is a 68% chance that the population mean will occur. This range is called a 68% **confidence interval**. However, this is a bit weak; we usually want to be rather more confident than only 68%. The greater the chance of our range containing the population mean, the wider the confidence interval must be.

Think about it: if we want to be absolutely certain that the range contains the population mean we need a 100% confidence interval and this must be very wide indeed; it must embrace the whole of the distribution curve not just the middle part. It is standard practice to calculate a 95% confidence interval which is usually taken as a good compromise. The 68% confidence interval was obtained by multiplying the standard error by one; therefore to calculate the 95% confidence interval we need a 'multiplier' of greater than one. Statistical tables, summarizing the Normal distribution curve, tell us that we need to multiply by 1.96 for a 95% confidence interval. There is however just one further modification to be made. The Normal distribution assumes that we have a large number of observations (usually taken to be more than 30) in our sample. If, as in our case, we have only ten our knowledge is more limited so we need to take this fact into account.

Having less information leads to more uncertainty and so to a wider confidence interval. A distribution which is appropriate for small samples is the **'t' distribution**. Again this exists in table form, so from the tables we can find out the appropriate value of 't' with which to multiply the standard error in order to obtain a confidence interval. The value of 't' increases as the number of observations decreases – there is more uncertainty and, as with the Normal distribution, the value of 't' increases as the amount of confidence we require increases. So, for example, with 30 worms in the sample and only 95% confidence required 't' = 2.04, but if 99% confidence is needed and the sample is only ten worms 't' increases to 3.35.

Degrees of freedom	95% Confidence	99% Confidence
9 (= 10 worms)		
9	2.26	3.35
19 (= 20 worms)		
19	2.09	2.84
29 (= 30 worms)		
29	2.04*	2.76

* This is very close to the value of 1.96 which comes from the Normal distribution. For most practical purposes, if our sample contains 30 or more individuals a 95% confidence interval is given by the mean ± twice its standard error.

2.1.1 So what's so important about a confidence interval?

The statement that 'the mean length was 94 cm' has some value, but not a great deal. It indicates that we are probably not referring to whales, but it doesn't give us any idea about the variability around that mean. It is this variation which is essential information if we are to take decisions.

For example, imagine that you are the owner of a factory which prepares and packages a special herbal tea which is in great demand. The distinctive element in the product is an extract from a fruit which is produced on a particular species of shrub in the local forest and which is harvested by hand. This forest is remote and its species composition is not well known. Market research indicates that the demand for the product is likely to continue to grow and you would happily invest several million pounds in expanding the factory, but you would only do this if you were confident that there was enough of the shrub present within a certain distance from the factory to keep a new and larger factory fully supplied in a sustainable way. If you decide to invest the money and then run out of fruit you will go bankrupt and the local people will lose their jobs. The way to gain information is to take a sample (Chapter 3).

The results from the sample survey are that there is a mean of 20 bushes per hectare. If every single bush (i.e. the population) had been found we would know that 20 bushes/ha was the true value. However, as we have only **sampled** some of the population we have only an estimate of the mean and we know that there is some uncertainty about how close the true population mean is to this estimate. If we calculate a confidence interval, we can quantify this uncertainty. The 95% confidence interval might be from 15 to 25 bushes per hectare (Fig. 2.1). So we are 95% confident that the population mean lies in this range. If we have worked out that, to make full use of the factory's capacity, we must have at least 15 bushes/ha we might decide that we are happy to proceed since there is only a small chance (2.5%, which is the area in the left-hand tail of the distribution) that the population mean will be less than this.

However, we may decide that a 2.5% chance of bankruptcy from this source of risk is too much and that we are only prepared to take a 0.5% risk. Then a 99% confidence interval for the population mean should be used instead. This gives the range which is 99% likely to include the population mean. This might be between 12 and 28 bushes. Then, there is a 0.5% chance of the population mean being less than 12 bushes per hectare (the 0.5% chance of there being more than 28 bushes per hectare does not influence the decision). This confidence interval now includes uneconomic densities of shrubs (12–14.9). Thus we may decide to leave the factory as it is.

In the real world there are likely to be areas of the forest where the bushes are dense and other areas where they are sparse. The mean of 20 bushes per hectare might be built up from a few areas at a great distance from the factory, where there are many shrubs and areas nearby where there are few. This would be vital additional information since it would affect the cost of collecting the fruit. In Chapter 3 we look at how sampling methods can be made more sophisticated to take account of this type of variation.

2.1.2 Consolidation of the basic ideas

You have now encountered some very important concepts. Don't be surprised if you don't understand or remember them after just one reading. Most people find that they need to come across these ideas many times before they are comfortable with them. The best way to understand the subject is to have some observations of your own which you wish to summarize, then you are motivated to put these methods to good use. In the mean time, use the flow chart below to obtain the standard error from the data which follow it. Check that you understand what is going on and why. If you are puzzled, refer back to the appropriate section and re-read it. Reading about statistical methods

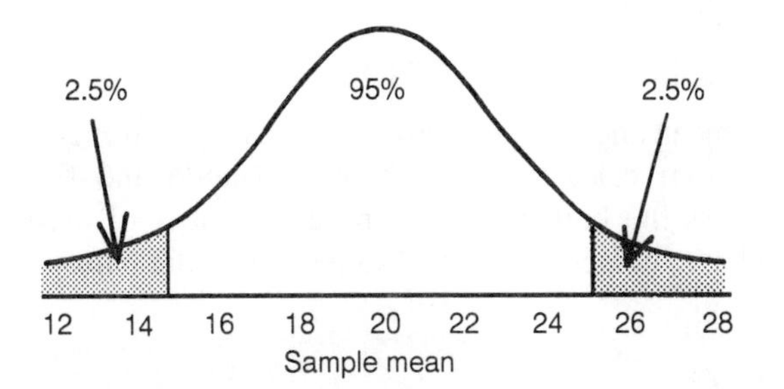

Figure 2.1 There is a 95% chance that this range includes the mean number of bushes per hectare over the entire forest.

cannot be done at the same fast pace as reading a light novel or a science fiction book – but it will repay your attention.

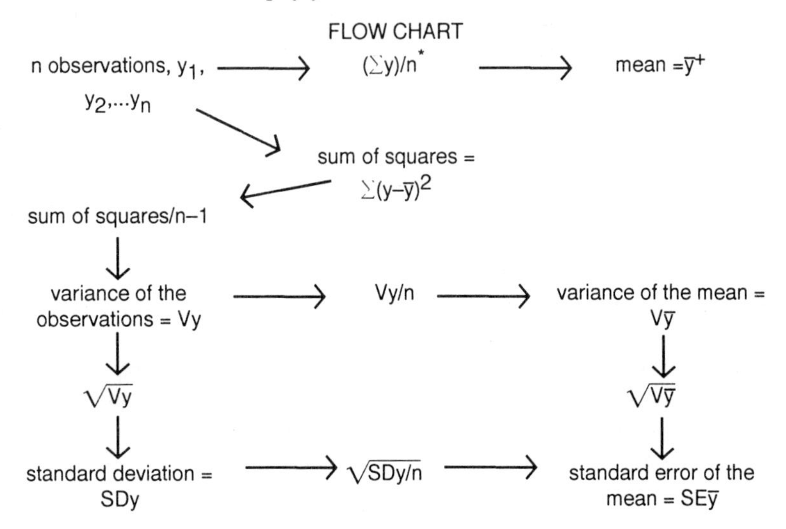

* The symbol $\sum$ means 'add up all the values of the item which follows'. In this case: all the n values of y.
+ The bar above y indicates that this is the mean of all the values of y.

2.1.3 Using the statistical mode on your calculator

Many calculators have a statistical mode. This allows us to short-circuit the above pathway by entering the data and obtaining the standard deviation directly by pressing the appropriate key usually either a button marked 's' or one marked:

$\sigma_{n\text{-}1}$

(depending on the make of the calculator).

Try out your calculator using this sample from a medical doctor's records of the ages (years) at which 12 people developed the symptoms of a particular disease:

27 32 36 38 45 48 55 61 68 71 75 78

To calculate the mean and a 95% confidence interval go through the following procedure. Put your calculator into statistical mode and then enter the observations. Press the button marked n (the number of observations) to check that you have, in fact, entered 12 values. Press the button marked:

$\bar{x}$

to obtain the mean.

To obtain the standard error we need to divide the standard deviation by the square root of n (the number of observations). Therefore we next calculate the square root of 12 and store it in the memory. Now press the standard deviation button and divide the result by the contents of the memory to give the standard error. Multiply the standard error by the value of t for 11 degrees of freedom (n-1) from the tables (95%) which is 2.201. Add the result to and subtract it from the mean to give the required confidence interval.

Here are the results: mean = 52.8, standard deviation = 17.6986, square root of 12 = 3.4641, standard error = 5.1092, t = 2.201, confidence interval = 52.8 plus and minus 11.245. From 41.5 to 64.0 years. We conclude that there is a 95% chance of the population mean lying within this range. What would be the range for a 99% confidence interval? Remember however that the population consists of the patients of one particular doctor. It would be misleading to generalize from these results to those patients on the list of a doctor in a different part of the country. No matter how good the statistical summary your conclusions must always be reviewed in the light of how the observations were collected originally.

2.2 INTRODUCTION TO MINITAB

We can improve on the calculator by using a computer package. The MINITAB package is available on PCs and is very popular. The latest version (Release 8) allows you to select commands from menus instead of typing them in. It is very easy to use and provides quick and useful ways of looking at the sample of observations or set of **data** we have collected and of summarizing it. We will see how it could be used to deal with our original sample of the lengths of ten worms. We type the ten values directly into column 1 of the worksheet and name the column 'length'. Then we can ask for the data to be printed:

```
MTB > print c1
length

6.0   8.0   9.0   9.5   9.5   10.0   10.5   11.5   12.5   13.5
```

Because there are only a few values they are printed across the page instead of in a column, to save space.

We can see the position of each observation on a scale of length:

```
MTB > dotplot c1
                                  :
       .          .         .     :    .    .          .    .          .
--------+------------+------------+------------+------------+------------+ --------length
       6.0          7.5          9.0         10.5         12.0         13.5
```

This shows us the length of each worm very clearly.

We can group the observations into classes (or groups) by length:

```
MTB > histogram c1
Histogram of length   N = 10
 Midpoint      Count
     6           1        *
     7           0
     8           1        *
     9           1        *
    10           3        ***
    11           1        *
    12           1        *
    13           1        *
    14           1        *
```

MINITAB produces histograms on their side (you may be more used to seeing them with the midpoints of the classes running along the bottom). Each star or asterisk represents one worm. The class with a midpoint of 8 contains any worm with a length of from 7.5 to 8.4 cm. In this case there is only one. There are no worms in our sample with lengths in the range from 6.5 to 7.4 cm, but there are three in the range from 9.5 to 10.4 cm.

A more detailed version of the above histogram is provided by the stem-and-leaf plot:

```
MTB > stem-and-leaf c1
Stem-and-leaf of length   N = 10
Leaf Unit = 0.10
   1        6      0
   1        7
   2        8      0
   5        9      055
   5       10      05
   3       11      5
   2       12      5
   1       13      5
```

This has three columns of numbers. The central column is called the stem and the numbers to its right are the leaves. There is one leaf for each observation. Here it represents the value of the number after the decimal point in the observation. There can be many leaves on each stem. The stem here represents the whole number part of an observation, before the decimal point. The top row of the middle and right-hand columns shows that there is one observation with a stem of 6 and a leaf of 0 (this is equivalent to a worm 6.0 cm in length). The fourth row shows that there is one worm 9.0 cm long and two worms are each 9.5 cm long.

The column of numbers on the left is the *cumulative* number of worms up to the mid-point of the histogram. Thus starting from the top these are 1 (one worm in first class), 1 again (no more worms added from second class), 2 (one more worm added – length 8.0 cm) and 5 (three more worms added in fourth class). The same process goes on from the bottom upwards as well, so that we can see that there are three worms whose lengths are greater than or equal to 11.5 cm and five worms with lengths of at least 10 cm. In our sample the observations happen to fall into two groups of five counting from both the high end and from the low end, but the distributions in small samples are not always quite so well balanced.

The advantage of the stem-and-leaf plot is that, in addition to giving us the shape of the distribution, the actual values are retained. The plot would also suggest to us (if we did not know already) that the lengths were recorded only to the nearest 0.5 cm which might affect how we interpret the results or the instructions we give to anyone who wanted to repeat the survey.

The following instruction produces many **summary statistics**.

```
MTB > describe c1

              N     MEAN    MEDIAN    TRMEAN    STDEV    SEMEAN
length       10   10.000     9.750    10.063    2.173     0.687

            MIN      MAX        Q1        Q3
length     6.00   13.500     8.750    11.750
```

N is the number of observations (the same as 'n' used in the text previously) and is followed by the **MEAN**. Further on there are minimum (MIN) and maximum (MAX) values. Notice that, in addition, MINITAB gives a 'trimmed mean' (TRMEAN). This excludes the smallest 5% and largest 5% of the values (rounded to the nearest whole number) before calculating the mean. Here, since 5% of a sample of ten is 0.5, one value has been excluded from each end. If there were a big difference between the mean and the trimmed mean it would suggest that one very large (or very small) value was greatly affecting the mean value. You might at this stage wish to review such values, to double-check that they have not been wrongly recorded or wrongly entered. Perhaps they may have some other characteristics that mean that they should be excluded from the sample. You might find on checking, for example, that the very long worm was, in fact, a different species.

The values headed 'STDEV' and 'SEMEAN' are the standard deviation and the standard error of the mean. These are the same as we obtained by laborious calculation earlier. That leaves us with three other statistics: 'MEDIAN', 'Q1' and 'Q3'. These belong to a slightly different way of summarizing data which is particularly useful when observations do not follow a Normal distribution. In such circumstances the mean is still an estimate of the average value but it may be a misleading one to use. For example, in most organizations there is a hierarchy of salaries from the most junior employee to the most senior one. Usually there are many more people earning relatively low salaries than high ones but if we calculate the mean salary it is greatly affected by the very high salaries of just a few individuals. The mean is then not the best measure of the salary of a 'typical' employee. Let's look at some actual figures. The salaries of 13 employees have been entered into column 1 of a MINITAB worksheet and printed out, as for the worm lengths, as follows:

```
MTB > print c1
salary (pounds per year)

  8500    9000    9500  10 000  10 000  10 500  11 500  12 000  12 000
18 000  17 500  25 000  54 000
```

Then a stem-and-leaf plot is produced:

```
MTB > stem-and-leaf c1
Stem-and-leaf of salary  N = 13
Leaf unit = 1000

   3     0   899
  (6)    1   000122
   4     1   78
   2     2
   2     2   5
   1     3
   1     3
   1     4
   1     4
   1     5   4
```

MINITAB chooses an appropriate method of rounding for the data set, so that the shape of the distribution is clear. The classes are in £5000 blocks. The stem gives tens of thousands and each leaf represents £1000. Therefore the smallest three values appear by a stem of zero (no tens of thousands) as 8, 9 and 9. The next line contains the six values between £10 000 and £14 999.

The **median** value is that value of salary which is exceeded by 50% of individuals. It is a good way of estimating the average salary for this data because it is not unduly affected by the few very high values. We know that the median is contained in the second class because the left-hand column shows the number of observations in that class (6) in parentheses.

Starting from the high end of the range we see that one person has a salary of £54 000 and then there is a long gap before there is another one who earns £25 000, making two cumulatively who earn £25 000 or more, whereas most people earn far less.

The mean salary is £15 952. If we use the trimmed mean this is reduced to £13 182 reflecting the influence of the one very high salary. However, the observations are not symmetrical around the mean but have much wider spread (tail) of high values than of low ones. This is called positive

skewness. If there were more high values than low ones this would be called negative skewness. In our case the median is a much better summary statistic than is the mean.

MTB > describe c1

	N	MEAN	MEDIAN	TRMEAN	STDEV	SEMEAN
salary	13	15 962	11 500	13 182	12 360	3428

	MIN	MAX	Q1	Q3
salary	8500	54 000	9750	17 750

The 'describe' command tells us that the median is £11 500. Q1 stands for **lower quartile**. This is the value which has one quarter (25%) of observations below it. Obviously some rounding is necessary where the number of observations is not exactly divisible by four. Similarly Q3 is the salary which is exceeded by one quarter (25%) of people. So three-quarters of people have salaries below it. It is called the **upper quartile**. MINITAB produces a very helpful graph which contains these features, called a box-and-whisker plot (or boxplot).

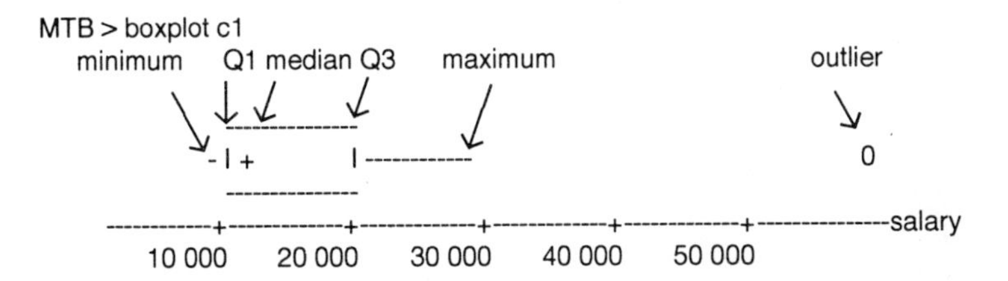

The cross shows the median. The box around it encloses the central 50% of the observations and so it excludes the largest 25% and smallest 25% of values. It is defined at the lower end by Q1 and at the upper end by Q2. The lines from each side of the box (called whiskers) extend as far as the minimum and maximum values except that very small or very big values (**outliers**) are marked separately beyond the ends of the lines as '*' or '0' to draw attention to them. Thus the £54 000 point is so marked in our example. This also makes it very easy to spot mistakes made in entering the data (for example if an extra zero had been added onto a salary of £10 000).

Contrast the boxplot for the salaries data with that for the lengths of the ten worms (cm) we measured earlier:

6.0, 8.0, 9.0, 9.5, 9.5, 10.0, 10.5, 11.5, 12.5, 13.5

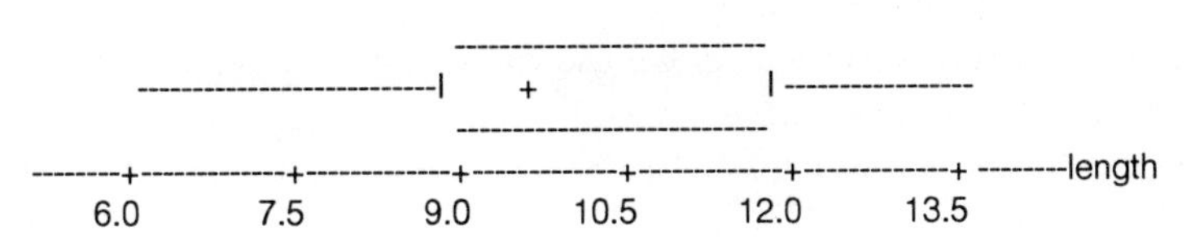

Here the median is nearer the middle of the box and the whiskers are symmetrical. It is often helpful to display the data in this way early on in the proceedings, to get an idea as to whether they do show a roughly Normal distribution (as for worm lengths) or not (salaries), because some of the tests that will be discussed later are based upon the assumption that the data are normally distributed.

Sampling

3

3.1 FIRST, CATCH YOUR WORM

In the last chapter I cheated and said 'let's imagine that we have collected ten worms'. Not a word about how we would do this, but we must now consider this challenge.

First, there are practical problems. If we go out during the day the worms will be buried and we will need to dig them up. How deep should we go? Also, we run the risk of cutting some of them by accident. It would probably be better to go out on a wet night with a torch and collect them into a bucket while they are squirming on the surface.

Second, how do we decide which ten worms to pick up from the many thousands available? If we collect them from near the footpath, because this is convenient and we won't get our shoes too dirty, we will have individuals which tend to be shorter than those in the rest of the field because that was where they happened to be infected by parasites. We will have failed to obtain a **sample** of ten worms which is **representative** of the **population** of thousands of worms in the field. So deciding which ten worms to pick is important and the process is called '**sampling**'. This chapter will describe three alternative methods of sampling (random, stratified random and systematic), together with their strengths and weaknesses.

3.2 RANDOM SAMPLING

The purpose of random sampling is to ensure that each worm in the population we are studying has an equal chance of being selected. If we have been successful in taking a random sample then it will provide an unbiased estimate of the mean. In other words if you took all possible random samples, each of ten worms, from the field, the sample means would vary – some would be bigger than the unknown population mean and some would be smaller. However, there would be as many 'high' values as 'low' ones and if you took the average of all of their means, that value would be the true population mean.

Unfortunately, if we usually take only one random sample of, say, ten observations we may obtain an **unrepresentative** result because, by chance, no observations have been selected from a particularly distinctive part of the population. Nevertheless if we start off knowing nothing about the worms and how they vary in length and abundance throughout the field a random sample is more likely to give us a representative sample than is any other approach.

3.2.1 How to select a random sample

We have decided that our population consists of the worms on the surface of a field on a damp night. Note that this immediately rules out extrapolating our results to worms that have stayed below ground. If we are about to set out on a sampling scheme we should think about whether our method automatically precludes some of the people, species, individuals or events in which we may be interested. Telephone surveys exclude any household that does not have a working telephone; surveys of woodland plants will under-record plants such as celandines and anemones if they take place in the late summer or autumn, after these species have died back. But back to the worms . . .

We can make a random sample of surface-crawling worms by selecting ten positions in the field at random, going to these and then picking up the nearest worm. Make a rough plan of the field and mark the length and width of it in metres (Fig. 3.1). Then press the random number button on the calculator (RAN #) to give the number of metres across the field and press the random number button again to give the number of metres down the field that we should go to find a sample point. For example, if the numbers generated are: 0.034 and 0.548 these can be interpreted as 34 m across and 48 m up (using the last two digits). Alternatively, we can use the first two values (like 34 and 48) from a table of random numbers, ignoring any which are too large.

Make sure that the range of coordinates which are available will allow any point in the field to be selected. If the field was more than 100 m in one direction for example at least three digits would be needed. Having found one point we repeat this process until we have found ten points which lie within the field. Let's look at a practical example which for simplicity will be a small square field, 400 m^2 in size. Figure 3.2 shows the yield of fruit from a particular shrub (g) on each 1 m^2 patch, conveniently arranged in 20 rows and 20 columns.

To select a position at random we could use the last two digits in the numbers generated by the random number button on our calculator. If this gives 0.302 and then 0.420 this would identify the position in row 2 and

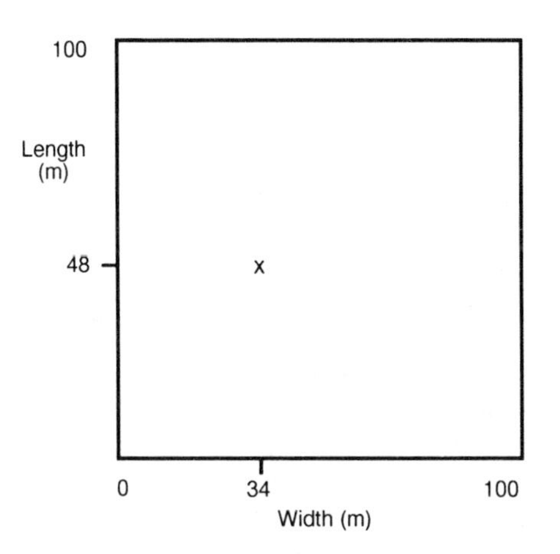

Figure 3.1 Position of a random sample.

column 20. If we select a random sample of 16 positions at which to harvest the crop and measure its yield we could get the selection shown in Fig. 3.2. One of the most common problems in sampling is deciding on how many observations should be made. A general rule is that the more variable is the population, the more observations must be taken from it. This is necessary if we are to obtain a 95% confidence interval which is not so large as to be useless. Here we have taken 16 observations. This could be regarded as a pilot study. If the confidence interval turns out to be very wide we will take more observations. At least we will be unlikely to have taken an excessive number of observations.

The standard error of the mean is then obtained by putting all the observations into the calculator on statistical mode, pressing the standard deviation button and dividing the answer by the square root of 16:

mean = 18.87g, SE mean = 2.979

We notice that by chance, in our sample of 16 points no positions were selected from the top right or bottom right of the field, where crop yields happen to be low. So this sample is unrepresentative although there is no way that, without some other information, this could have been predicted in advance. However, if we started off knowing which areas were particularly low and which high yielding then we could improve our sampling technique (section 3.3).

3.3 STRATIFIED RANDOM SAMPLING

Where there is existing information or good reason to believe that the variation in the feature we want to sample is not evenly distributed, then it is sensible to divide the population into sections called **strata** (singular

Row	COLUMN 1	2	3	4	5	6	7	8	9	10	11	12	13	14	15	16	17	18	19	20
1	8	7	12	9	6	8	12	14	7	4	5	4	3	7	4	2	4	6	8	7
2	16	18	9	10	7	9	14	10	6	3	4	8	5	6	5	3	3	4	6	6
3	22	**15**	17	14	12	13	16	9	4	5	6	5	7	7	6	5	4	5	7	9
4	19	16	13	10	8	9	12	9	8	7	9	10	9	8	3	3	7	10	13	10
5	16	12	12	9	5	7	10	12	6	8	7	6	6	5	6	4	6	7	9	10
6	19	16	11	7	8	**10**	14	10	16	15	17	16	18	17	19	22	24	20	19	15
7	22	18	16	19	12	13	17	12	18	**21**	20	19	25	22	24	27	**20**	17	22	14
8	25	14	16	12	13	15	16	14	23	25	28	33	30	27	31	25	22	20	24	29
9	20	17	14	**16**	17	19	**20**	24	25	27	32	40	42	35	35	37	29	28	27	**35**
10	**25**	22	19	25	29	32	36	30	35	40	**41**	47	45	40	32	33	27	27	31	30
11	29	30	27	32	37	42	45	45	43	51	53	52	48	39	42	**37**	33	29	27	20
12	31	35	38	45	47	44	49	51	**50**	58	56	50	41	43	30	27	29	23	21	**19**
13	28	33	30	**37**	40	39	42	46	51	48	44	40	37	30	35	22	21	17	14	16
14	26	29	32	35	35	31	29	32	40	37	35	31	16	19	20	11	12	10	13	12
15	22	28	31	29	28	27	25	27	16	22	19	16	8	4	2	3	6	5	7	10
16	21	22	28	26	31	27	22	24	**20**	14	10	8	6	6	4	7	4	3	1	5
17	17	26	30	31	31	**27**	25	33	12	8	7	6	6	4	2	1	0	0	0	0
18	17	24	31	33	27	22	19	17	14	10	9	7	3	0	3	1	0	4	2	0
19	19	19	21	17	15	16	16	19	16	13	10	12	8	3	1	0	3	2	5	3
20	23	**21**	17	14	13	19	23	17	12	11	6	9	7	3	1	0	1	1	2	4

Figure 3.2 Random sample of 16 patches

stratum). For example, if we could see that the field in the example above has a slope, with a stream at the bottom, we could make one stratum at the top, two in the middle and one at the bottom of the slope. This would be likely to give four strata which have soils with different moisture contents, but with a reasonably consistent amount of soil moisture within a particular stratum (conditions are **homogeneous** within it) and we might expect that soil moisture is likely to influence yield. Also, although we wouldn't be able to see all the crop yields in real life we might well notice that the shrubs are biggest in the central band of the slope and especially small in the top right and top left of the field. We therefore could split each of the four horizontal strata in two down the middle to form eight strata – with the aim of having roughly even shrub size within any one of them. This is called **stratification** (Fig. 3.3).

We have successfully identified a pattern in the population and formed different sub-populations which will enable us to improve the effectiveness of our sampling of the yield of the shrubs in the field.

Stratifying in the opposite direction in this case would not be sensible. The stratification system we use should be based on features that are important for what we want to observe or measure, not on arbitrary lines or in ways that do not reflect the underlying variation. For example we could have divided up our field as in Fig. 3.4. This would lead to some

ROW	COLUMN 1	2	3	4	5	6	7	8	9	10	11	12	13	14	15	16	17	18	19	20
1	8	7	12	9	6	8	12	14	7	4	5	4	3	7	4	2	4	6	8	7
2	16	18	9	10	7	9	14	10	6	3	4	8	5	6	5	3	3	4	6	6
3	22	15	17	14	12	13	16	9	4	5	6	5	7	7	6	5	4	5	7	9
4	19	16	13	10	8	9	12	9	8	7	9	10	9	8	3	3	7	10	13	10
5	16	12	12	9	5	7	10	12	6	8	7	6	6	5	6	4	6	7	9	10
6	19	16	11	7	8	10	14	10	16	15	17	16	18	17	19	22	24	20	19	15
7	22	18	16	19	12	13	17	12	18	21	20	19	25	22	24	27	20	17	22	14
8	25	14	16	12	13	15	16	14	23	25	28	33	30	27	31	25	22	20	24	29
9	20	17	14	16	17	19	20	24	25	27	32	40	42	35	35	37	29	28	27	35
10	25	22	19	25	29	32	36	30	35	40	41	47	45	40	32	33	27	27	31	30
11	29	30	27	32	37	42	45	45	43	51	53	52	48	39	42	37	33	29	27	20
12	31	35	38	45	47	44	49	51	50	58	56	50	41	43	30	27	29	23	21	19
13	28	33	30	37	40	39	42	46	51	48	44	40	37	30	35	22	21	17	14	16
14	26	29	32	35	35	31	29	32	40	37	35	31	16	19	20	11	12	10	13	12
15	22	28	31	29	28	27	25	27	16	22	19	16	8	4	2	3	6	5	7	10
16	21	22	28	26	31	27	22	24	20	14	10	8	6	6	4	7	4	3	1	5
17	17	26	30	31	31	27	25	33	12	8	7	6	6	4	2	1	0	0	0	0
18	17	24	31	33	27	22	19	17	14	10	9	7	3	0	3	1	0	4	2	0
19	19	19	21	17	15	16	16	19	16	13	10	12	8	3	1	0	3	2	5	3
20	23	21	17	14	13	19	23	17	12	11	6	9	7	3	1	0	1	1	2	4

Figure 3.3 Eight strata

strata having a mixture of big and small shrubs and is unlikely to improve the effectiveness of our sampling if (as is likely) big and small shrubs differ considerably in their yields.

When the boundaries of the eight strata have been fixed, then two (or more) sample positions are selected at random from within each of them. This is called **stratified random sampling** (Fig. 3.5a). To select the points for each stratum you can use the same procedure as before except that once you have two points in a stratum you pass over any further ones that fall in it, and keep going until you have two points in each. Alternatively you may find it quicker to work out sets of coordinates for each stratum separately. For irregularly shaped strata this can be done by overlaying them with a grid of points (Fig. 3.5b). The grid must of course be larger than the whole area and the grid points outside the stratum are not used.

Here we are dealing with strata that are approximately equal in size. The method can be generalized to unequal-sized strata.

When the area has been stratified into several roughly equally-sized strata we use a method which differs slightly from that adopted for simple random sampling to obtain the standard error of the mean. The data for a given stratum are entered into the calculator which, as before, is on statistical mode. We then work out the variance of the mean for that

ROW \ COLUMN	1	2	3	4	5	6	7	8	9	10	11	12	13	14	15	16	17	18	19	20
1	8	7	12	9	6	8	12	14	7	4	5	4	3	7	4	2	4	6	8	7
2	16	18	9	10	7	9	14	10	6	3	4	8	5	6	5	3	3	4	6	6
3	22	15	17	14	12	13	16	9	4	5	6	5	7	7	6	5	4	5	7	9
4	19	16	13	10	8	9	12	9	8	7	9	10	9	8	3	3	7	10	13	10
5	16	12	12	9	5	7	10	12	6	8	7	6	6	5	6	4	6	7	9	10
6	19	16	11	7	8	10	14	10	16	15	17	16	18	17	19	22	24	20	19	15
7	22	18	16	19	12	13	17	12	18	21	20	19	25	22	24	27	20	17	22	14
8	25	14	16	12	13	15	16	14	23	25	28	33	30	27	31	25	22	20	24	29
9	20	17	14	16	17	19	20	24	25	27	32	40	42	35	35	37	29	28	27	35
10	25	22	19	25	29	32	36	30	35	40	41	47	45	40	32	33	27	27	31	30
11	29	30	27	32	37	42	45	45	43	51	53	52	48	39	42	37	33	29	27	20
12	31	35	38	45	47	44	49	51	50	58	56	50	41	43	30	27	29	23	21	19
13	28	33	30	37	40	39	42	46	51	48	44	40	37	30	35	22	21	17	14	16
14	26	29	32	35	35	31	29	32	40	37	35	31	16	19	20	11	12	10	13	12
15	22	28	31	29	28	27	25	27	16	22	19	16	8	4	2	3	6	5	7	10
16	21	22	28	26	31	27	22	24	20	14	10	8	6	6	4	7	4	3	1	5
17	17	26	30	31	31	27	25	33	12	8	7	6	6	4	2	1	0	0	0	0
18	17	24	31	33	27	22	19	17	14	10	9	7	3	0	3	1	0	4	2	0
19	19	19	21	17	15	16	16	19	16	13	10	12	8	3	1	0	3	2	5	3
20	23	21	17	14	13	19	23	17	12	11	6	9	7	3	1	0	1	1	2	4

Figure 3.4 Poor stratification

stratum by calculating the standard deviation from our calculator button (s or σ_{n-1}), squaring it (a quick way to return to the variance) and dividing by the number of observations, (n = 2) in the stratum. (See the flow chart at 2.1.2.) Let's calculate this for Fig. 3.5:

Observations g/m^2		Stratum mean	Variance of stratum mean
9	13	11.0	4.0*
12	14	13.0	1.0
38	43	40.5	6.25
19	26	22.5	12.25
6	8	7.0	1.0
24	29	26.5	6.25
21	30	25.5	20.25
4	6	5.0	1.0
Total = 302		151.0	52.0

* e.g. standard deviation = 2.8284 (from calculator), variance = $(2.8284)^2 = 8.0$, variance of the mean for the stratum = 8.0/2 = 4.0.

The sum of the eight stratum means is 151 which when divided by eight gives the sample mean = 18.87, the same as for a random sample with the same observations. What, however, is its standard error?

(a)

Row \ COLUMN	1	2	3	4	5	6	7	8	9	10	11	12	13	14	15	16	17	18	19	20
1	8	7	12	9	6	8	12	14	7	4	5	4	3	7	4	2	4	6	**8**	7
2	16	18	9	10	7	**9**	14	10	6	3	4	8	5	**6**	5	3	3	4	6	6
3	22	15	17	14	12	13	16	9	4	5	6	5	7	7	6	5	4	5	7	9
4	19	16	**13**	10	8	9	12	9	8	7	9	10	9	8	3	3	7	10	13	10
5	16	12	12	9	5	7	10	12	6	8	7	6	6	5	6	4	6	7	9	10
6	19	16	11	7	8	10	14	10	16	15	17	16	18	17	19	22	24	20	19	15
7	22	18	16	19	12	13	17	12	18	21	20	19	25	22	24	27	20	17	22	14
8	25	14	16	**12**	13	15	16	14	23	25	28	33	30	27	31	25	22	20	**24**	29
9	20	17	**14**	16	17	19	20	24	25	27	32	40	42	35	35	37	**29**	28	27	35
10	25	22	19	25	29	32	36	30	35	40	41	47	45	40	32	33	27	27	31	30
11	29	30	27	32	37	42	45	45	**43**	51	53	52	48	39	42	37	33	29	27	20
12	31	35	**38**	45	47	44	49	51	50	58	56	50	41	43	**30**	27	29	23	**21**	19
13	28	33	30	37	40	39	42	46	51	48	44	40	37	30	35	22	21	17	14	16
14	26	29	32	35	35	31	29	32	40	37	35	31	16	19	20	11	12	10	13	12
15	22	28	31	29	28	27	25	27	16	22	19	16	8	4	2	3	6	5	7	10
16	21	22	28	26	31	27	22	24	20	14	10	8	6	6	4	7	4	3	1	5
17	17	**26**	30	31	31	27	25	33	12	8	7	6	**6**	4	2	1	0	0	0	0
18	17	24	31	33	27	22	19	17	14	10	9	7	3	0	3	1	0	**4**	2	0
19	19	19	21	17	15	16	16	**19**	16	13	10	12	8	3	1	0	3	2	5	3
20	23	21	17	14	13	19	23	17	12	11	6	9	7	3	1	0	1	1	2	4

b)

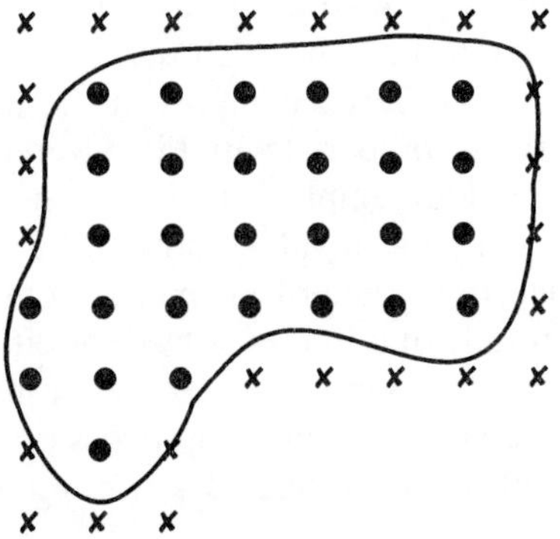

Figure 3.5 (a) Stratified random sampling of two patches in each of eight strata. (b) Select random points from those within the area marked '•'.

The sum of the variances of the stratum means is 52. There are eight strata but to obtain the variance of the overall mean we divide not by 8 but by 8^2 which is 64. This gives 0.8125. It may seem odd to divide the sum of the eight variances of the stratum means by 8^2 instead of just by 8 as we did for the mean. This is because the variance is in squared units, whereas the mean is in ordinary units. The standard error of the mean is then found by taking the square root of the variance of the mean to give 0.9014. This is much smaller than the standard error of the mean from our random sample (3.768).

So we have improved our estimate of the mean since the 95% confidence interval derived by multiplying the standard error by an appropriate value of 't' will be smaller. For example we might now be able to state with a particular degree of confidence that the population mean lies within the range from 17.9 to 19.9 instead of from 16.9 to 20.9 derived from the random sample.

The value of 't' for a stratified random sample with two samples per stratum is found for degrees of freedom equal to the number of stratum. Here this is 8 degrees of freedom. This is because we add together the degrees of freedom for each stratum $(1 + 1 + 1 + 1 + 1 + 1 + 1 + 1 = 8)$ just as we added together the estimates of variability from within each stratum (above).

A stratified random sample which has been sensibly allocated to a population with some pattern in it will always give rise to a smaller standard error than that from a random sample. This is defined as improving the **precision** of our estimate. Why does this happen?

Stratified random sampling splits the total variation in the sample into two parts: that caused by variation **within** each stratum and that caused by variation **between** the strata. The latter is then excluded and the standard error is calculated from within-stratum variation only. This should be relatively small, since strata have been chosen to be internally homogeneous. In other words the differences between the observations in each stratum are relatively small compared with the differences between **all** the observations in a fully random sample. That is why the stratification in Fig. 3.4 is considered poor – the mean value may still be the same/similar but the variance and hence the standard error and confidence interval are larger. If strata are naturally of different sizes or of different variability we should make more observations in the bigger and/or more variable strata.

In summary we can think of random sampling as providing an insurance policy against bias and of stratification as a way of improving precision.

3.4 SYSTEMATIC SAMPLING

Another form of sampling that you may come across is systematic sampling. In systematic sampling sample units are chosen to achieve maximum dispersion over the population. They are not chosen at random but regularly spaced in the form of a grid (Fig. 3.6a).

Systematic sampling is very efficient for **detecting** events because we are

(a)

Row	COLUMN 1	2	3	4	5	6	7	8	9	10	11	12	13	14	15	16	17	18	19	20
1	8	7	12	9	6	8	12	14	7	4	5	4	3	7	4	2	4	6	8	7
2	16	18	9	10	**7**	9	14	10	6	**3**	4	8	5	6	**5**	3	3	4	6	**6**
3	22	15	17	14	12	13	16	9	4	5	6	5	7	7	6	5	4	5	7	9
4	19	16	13	10	8	9	12	9	8	7	9	10	9	8	3	3	7	10	13	10
5	16	12	12	9	5	7	10	12	6	8	7	6	6	5	6	4	6	7	9	10
6	19	16	11	7	8	10	14	10	16	15	17	16	18	17	19	22	24	20	19	15
7	22	18	16	19	**12**	13	17	12	18	**21**	20	19	25	22	**24**	27	20	17	22	**14**
8	25	14	16	12	13	15	16	14	23	25	28	33	30	27	31	25	22	20	24	29
9	20	17	14	16	17	19	20	24	25	27	32	40	42	35	35	37	29	28	27	35
10	25	22	19	25	29	32	36	30	35	40	41	47	45	40	32	33	27	27	31	30
11	29	30	27	32	37	42	45	45	43	51	53	52	48	39	42	37	33	29	27	20
12	31	35	38	45	**47**	44	49	51	50	**58**	56	50	41	43	**30**	27	29	23	21	**19**
13	28	33	30	37	40	39	42	46	51	48	44	40	37	30	35	22	21	17	14	16
14	26	29	32	35	35	31	29	32	40	37	35	31	16	19	20	11	12	10	13	12
15	22	28	31	29	28	27	25	27	16	22	19	16	8	4	2	3	6	5	7	10
16	21	22	28	26	31	27	22	24	20	14	10	8	6	6	4	7	4	3	1	5
17	17	26	30	**31**	31	27	25	33	12	**8**	7	6	6	4	**2**	1	0	0	0	**0**
18	17	24	31	33	27	22	19	17	14	10	9	7	3	0	3	1	0	4	2	0
19	19	19	21	17	15	16	16	19	16	13	10	12	8	3	1	0	3	2	5	3
20	23	21	17	14	13	19	23	17	12	11	6	9	7	3	1	0	1	1	2	4

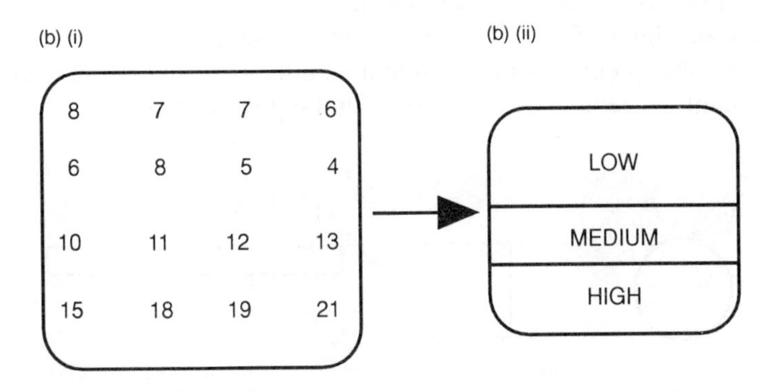

Figure 3.6 (a) Systematic sample of 16 patches. (b)(i) Observations from a systematic sample; (ii) stratified random sampling boundaries.

less likely to miss one (perhaps a fallen tree within a wood or a molehill in grassland) than if the sampling has a random element. Also, because you are more likely to include both very small and very large individuals, the mean of a homogeneous population is often close to the true mean, but has a large standard error.

If we have decided to use a systematic sampling grid then the first point should be chosen at random. Once this is chosen (for example, column 15, row 2 above) the other sample points are chosen in a fixed pattern from this point according to the scale of the grid and so they are not independent from each other. It is important to locate the first point at random because otherwise there is a risk that the grid will be positioned such that all the points miss (say) the edges of the sample area.

Much use is made of systematic sampling and the data are often treated as if they were from a random sample. For example in work in forestry plantations every tenth tree in every tenth row may be measured. As long as the number of sample units is high there is little risk of coinciding with any environmental pattern which might affect tree growth.

Similarly in social surveys, every 50th name on the electoral roll might be selected as a person to be interviewed. This is very convenient. However, it is important to be aware of the possibility of bias in systematic surveys. In the social survey for example the flats might be in blocks of 25 and all occupied by couples so we could end up only interviewing people who lived on the ground floor. In the forestry survey every tenth tree might coincide with the spacing of the forest drains so that all the sampled trees were growing a little bit better than their neighbours on the wet site.

Systematic sampling is excellent for mapping an unknown area however and for looking for patterns that you may wish to investigate in later samples. The yields of fruit per shrub taken from 16 trees distributed evenly as points in a grid can be used to divide the area into parts of low, medium and high productivity (Fig. 3.6b). Such parts could be used as strata in subsequent stratified random sampling, to obtain an unbiased, precise confidence interval for the population mean yield.

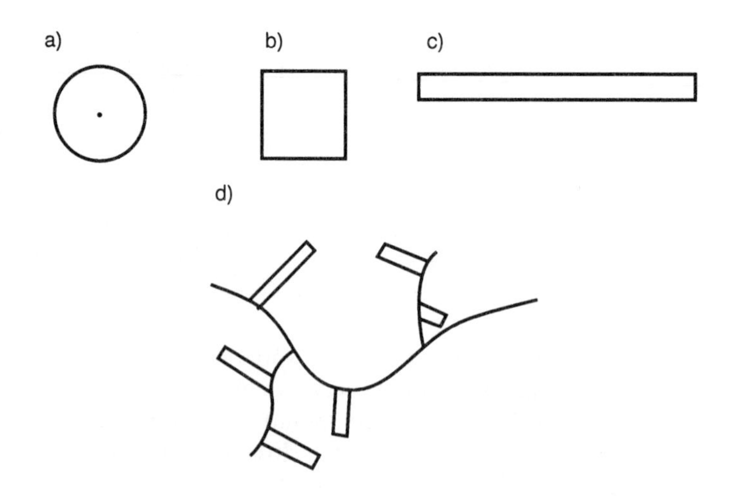

Figure 3.7 Types of sampling plot. (a) Circular; (b) square; (c) transect; (d) use of transects from paths.

3.4.1 Further methods of sampling

The basic methods of sampling may be made more sophisticated. For example, strata may naturally be of different sizes or we may choose to sample a very variable stratum more intensively. This is called 'irregular stratified random sampling'. We may select individuals in groups or 'clusters' for convenience by choosing several National Trust properties at random and then recording the height of all specimens of Wellingtonia trees at each.

Sampling may be 'two-stage'. Here we take a random sample of, for example, trees in a wood and then count the insects we find on a random sample of leaves on each of these trees. In addition, we might decide to select the trees with 'probability (of selection being) proportional to size', in other words we select so that a bigger tree stands a greater chance of being selected than a smaller one. This can be an efficient method of sampling because the population mean depends more on the means of the large trees than on those of the smaller ones. Sampling is a huge subject area in itself. As a beginner, you won't need to use any of these more sophisticated methods, but it is always a good idea to discuss your proposed method with someone who has experience of sampling. If nothing else they may be able to provide you with a few practical tips (for example, when doing field work never be separated from your waterproof clothing and your sandwiches!).

3.5 PRACTICAL PROBLEMS OF SAMPLING

In real life areas of land tend to be strange shapes rather than square. Then we must mark the boundaries of homogeneous areas as strata on a map. Draw a baseline along the bottom of the map (the x axis) and lines up the sides at right-angles to it (the y axis). Mark distances in metres along the x and y axes of the map. Then select the positions of the required number of small sampling areas (these are usually called **quadrats**) in each stratum using random numbers to identify the x and y grid coordinates.

In the field the quadrats can be sited in turn by pacing 1 m strides up and across, placing, say, the bottom left corner of the quadrat where the toe of the last foot falls.

We can make sample plots circular, square (rectangular) or in the shape of a long, thin strip (called a transect) (Fig. 3.7a–c). The advantage of a circular plot is that we need to mark its position only at one central point where we can then fix a tape measure rather than having to mark all the four corners of a square. We can use a transect when it is difficult to move through the vegetation. Strips cut into very dense vegetation at right-angles from tracks and positioned at random starting points along the paths are more efficient than random quadrats covering the same area, because we do not waste time in getting to them (Fig. 3.7d).

We must mark quadrats securely if we want to re-record them later. A couple of wooden pegs is not likely to be sufficient to survive several years of field wear and tear or vandalism. A buried metal marker (15 cm of thick metal piping) provides added security and we can re-locate it using a metal detector. If we intend to make observations at the same place at several dates a square quadrat is advantageous. It is easier to decide whether say, new seedlings are inside or outside it and, if there are large numbers of them, we can decide to sample only part of the quadrat (**subsampling**). For example, although we might record cover of established plant species over the whole quadrat, we might count seedlings present in the bottom right quarter only.

If we divide a 1 m square quadrat into, say, 25 squares each of 20 × 20 cm we can then record the presence or absence of particular species within each of these small squares. Then each species will have a frequency out of 25 for each quadrat. This is an objective method which encourages careful scrutiny of the whole quadrat and allows us to re-record sub-quadrats over time. It gives figures which are likely to be more reliable than subjective estimates of percentage ground cover for each species. Such subjective estimates of cover vary greatly from one observer to another. Also, the appearance of the same species under different managements or at different seasons may lead us to under- or over-estimate its importance.

Planning an experiment

4

So far we have learned how to estimate the mean of a large population by observing a small sample of individuals taken from it. Very often, however, we wish to compare the mean performance of several different categories. For example:

- which of three fertilizers gives the best crop yield?
- which of four species of worm moves the fastest?
- which drug cures the disease most rapidly?

In this chapter we will see how to design experiments which will answer such questions.

Let's start with a comparison of the yields of barley obtained from applying three increasing amounts of fertilizer (A, B and C where A is the lowest amount and C the highest). The term **treatment** is used to describe that which we are varying in our experiment. In the experiment that we are going to design we have three treatments.

4.1 REPLICATION

Our first thought might be just to split a field in three and apply a different amount of fertilizer to each third (Fig. 4.1a). This could give misleading results however. Suppose that the natural fertility of the soil is higher at the bottom of the slope, then whichever fertilizer is allocated to that position will appear to be better than it really is in comparison with the others. However, we could divide the field into, say, 12 parts or **experimental units** (usually called **plots**) and allocate each fertilizer treatment at random (see below) to four of them.

This will improve matters if there is variation in fertility in the field to start with because it is unlikely that all of one treatment will end up in a very high or very low fertility patch. Rather the underlying variation is likely to be spread between the treatments (Fig. 4.1b). We now have four replicates of each treatment.

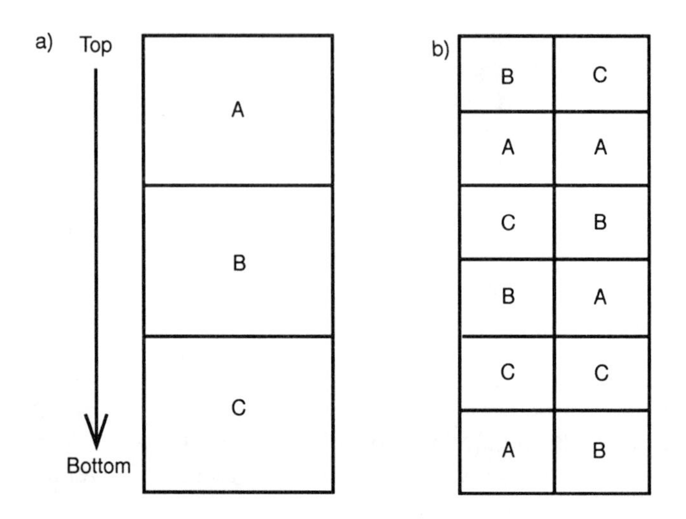

Figure 4.1 (a) No replication; (b) four replicates of each treatment.

We now compare the yields from just the plots that receive fertilizer treatment A. Any variation in the yields between the four plots is caused by random variation in conditions across the site. The same is true for variation in the four yields from treatment B and separately, from treatment C. So replication has given us three separate estimates of the background or random variation in the yields of plots receiving the same treatment. This provides a basis for comparison with the differences **between** the treatments that might be due to the fertilizer treatments. It is essential to have at least two replicates of each treatment but four replicates is commonly considered to be a minimum in field experiments.

The greater the replication we have (perhaps we could have six plots of each treatment instead of four), the better is our estimate of the random variation effects within the plots in that treatment. As in the previous chapter our estimate of the mean yield of treatment A, for example, is improved and the confidence intervals attached to the mean are reduced. Thus we should be able to detect smaller real differences in yield *between* the treatments from which we have taken our samples. We can achieve increased **precision** in our estimates of the population mean yields and of differences between them by having more replication.

4.2 RANDOMIZATION

The four replicate plots of each treatment must be allocated to positions in the field at random. This is achieved by numbering the 12 plots from 1 to 12 (Fig. 4.2a). Then we use the random number button on the calculator (or use random number tables) to select four of these plots for treatment A (e.g. using the last two digits from the numbers 0.809, 0.312, 0.707, 0.836 and 0.101 allocate this treatment to plot numbers 9, 12, 7 and 1 respectively; we ignore the value 36 from 0.836 because it is greater than 12) (Fig. 4.2b). Then we select four more numbers for treatment B and treatment C must go on thc four plots which rcmain.

As with selecting a sample from one population (Chapter 3) we can think of randomization in allocating treatments to experimental units as an insurance policy. It protects us from obtaining estimates of treatment means which are biased (consistently higher or lower than the population mean – say, because all plots from one treatment were in the corner of the field where the crop was damaged by frost). Randomization is also necessary because it helps to ensure that we can carry out a valid statistical analysis of the data (Chapter 6).

The replicates of each treatment must be **interspersed** (mixed up) over the site. Randomization is a good way of achieving this. Very rarely, randomization may produce an arrangement of plots in which say the four replicate plots of one treatment are grouped together at the bottom of the slope. Although this would be acceptable if this were one of a series of similar experiments, in real life this is unfortunately seldom the case. To avoid the possibility of our one-and-only experiment giving unrepresentative results, we could re-randomize the positions of the treatments, so that they are satisfactorily interspersed. However do not go to the other extreme of imposing a systematic layout of the plots such that treatment B

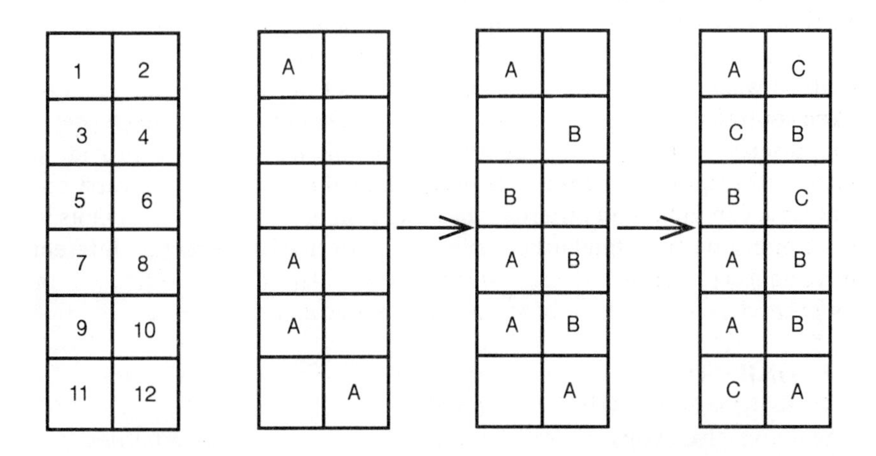

Figure 4.2 (a) Allocation of plot numbers; (b) allocation of treatments to plots using random numbers.

is always between A and C. This may also create problems (as well as being statistically invalid) if, for example, it means that B is always being shaded by the taller growing C.

4.3 CONTROLS

Suppose we carry out our fertilizer experiment and there are differences between the treatments with the yields from A being the least. How will we know that A is in fact having any effect on growth at all? What we need is a **control**. A control is the name given to a treatment in which (usually) nothing is applied to the plot. We can then see what changes take place naturally during the experiment. A slight variation on this idea is a procedural control. For example, suppose the experiment is taking place not in a field but in a glasshouse and the treatments are different chemicals applied in water. If we have a treatment where nothing is applied the results we get may be simply the effect of the water applied with chemicals. Therefore we might apply water only as a control. This allows us to assess any effect of the chemicals separately from that of the water.

So far we have been considering a simple experiment in which fertilizers are applied to barley. Imagine a more complicated one in which we are interested in the effect of sheep grazing at different times of year on the number of wildflower species in grassland at different sites across the country. We can fence off each plot to make sure that the sheep are in the right place at the right time each year for, say, 5 years. But the grassland present at the beginning will contain a range of species characteristic of its location, soil type and previous management. These may differ from one place to another, for example one site had always been grazed with sheep whereas another was growing maize 2 years ago.

If we want to carry out the same experiment on several sites, we must have control plots on each site. They tell us what happens on each site in the absence of any grazing treatments (in this instance) and provide a standard comparison between the two sites. In the same way if we carry out the same experiment in different years the results from the control plots in each year provide a standard basis for comparison of the effect of different treatments.

4.4 OBJECTIVES

The above gives you an idea of the key factors in experimental design – replication, randomization and controls, but there is an important stage we have skipped – precisely what are you trying to test in your experiment. It is always good practice to write down the background to your experiment.

This consists of why you are interested in the problem and your general objective. For example, farmers are being paid to sow grass in strips of land round the edges of their cornfields to encourage wildlife. The seed used to establish these strips could also contain wildflower seeds (but at much greater cost) which will attract butterflies and bees or spiders and may be generally thought to be better for nature conservation than strips without wild flowers. However the strips may also harbour weeds that could spread into the crop and reduce crop yield. How often and at what times of year should the grass strips be cut if we want to encourage the growth of the desirable wildflower species without increasing the competitive weed species. Is it possible to recommend the best solution? Let us take this question and see where it leads us in terms of trying to design and lay out an experiment, not forgetting the practical issues of how and what we record that will need to be faced before we ever get results to analyse.

4.5 LAYING OUT THE EXPERIMENT

Talking to farmers we have discovered that they think two cuts a year are needed to stop thistles flowering and setting seeds which then become weeds in the crop. However, wildlife advisers believe that two cuts may be too much for the sown wild flowers. Our first questions then become:

1. What is the effect of sowing or not sowing wild flowers?
2. What is the difference between one cut and two cuts?

We have decided to compare four treatments:

	Cutting once/year	Cutting twice/year
With flower seeds	F1	F2
No flower seeds	NF1	NF2

At this stage we might well also sensibly decide to include a control treatment. But, for simplicity of explanation here we will deal only with the above four treatments. We could decide to have four replicate lengths of strip (plots) for each of our treatments. We must decide the size of each strip. Its depth from the edge of the field to the crop will be as used by farmers (say, 10 m) but how long should a plot be? It is important to have enough room to turn mowing machines round at the ends of each plot. Also it's possible that wildflower seeds may blow a metre or so into a neighbouring plot at sowing.

When a particular plot has received its one cut of the year it will grow quite tall in the rest of the year and may shade the edge of the neighbouring plot which receives a further cut. All these facts point to the idea of having a **buffer zone** round the edge of each plot (Fig. 4.3a). This is treated as far as possible like the rest of the plot but we will not make any recordings from it because it is unlikely to be a good representation of the treatment because it may be affected by the neighbouring treatment. Its function is to protect the inside of the plot.

It is common to use a 1 m × 1 m quadrat for recording vegetation. There will probably be variation within each plot – perhaps caused by soil composition, shading by occasional large shrubs in the nearby hedge or rabbit grazing. Just as it was desirable to have replicates of each treatment, so it is desirable to take more than one sample from within each strip. We now decide how we are going to sample each particular treatment strip. We will take the mean value from several quadrats as a fairer representation of the plot as a whole than if we had recorded only one quadrat which happened by chance to have been on the regular pathway of a badger.

Choosing the number of replicates and samples within each replicate is a problem. On the one hand the more we have, the more precision we will have in our results. On the other hand we will have more work to do. We need to consider the cost and availability of resources: people, land, equipment and the time and effort involved. Tired people will start to make mistakes at the end of a hard day's work, no matter how dedicated they are. Somewhere there is a compromise to be reached. Let's assume that we decide to have three quadrats on each plot. This will allow us to decide on the length of the plot, after allowing room between each quadrat

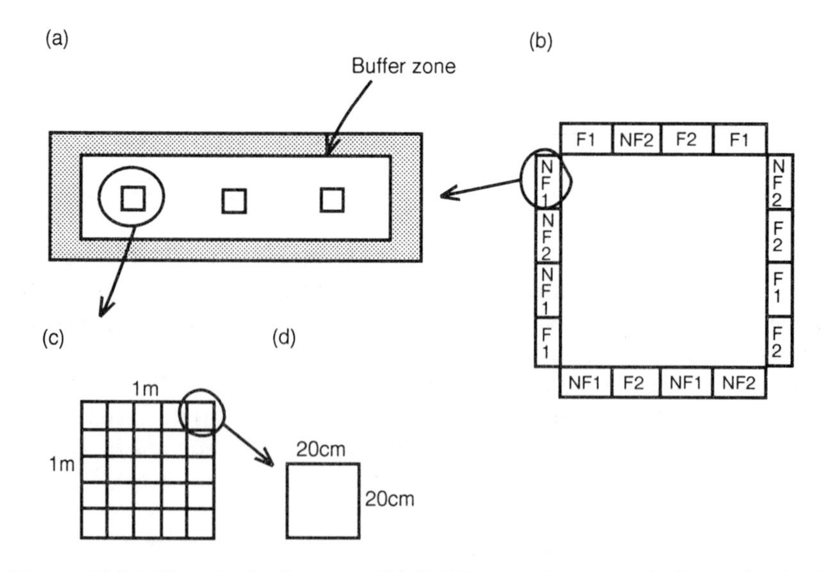

Figure 4.3 (a) Plot with buffer zone; (b) field layout of plots – wholly randomized design; (c) quadrat; (d) presence or absence recorded in each small square.

to move around and place equipment. We now have 16 × 3 = 48 quadrats to record in total.

Our four replicate plots of each treatment have been allocated at random to 16 positions round the field (Fig. 4.3b). They've all been ploughed and made level, with eight being sown to grass and eight to grass and wild flowers, as appropriate and we know that we will be recording three 1 m × 1 m quadrats in each plot. What, however, will we record in each?

4.6 RECORDING DATA

To monitor the effect of the treatments on species composition we could make a list of the species occurring in each quadrat. It is a good idea to do this as soon as the plants have established themselves and are recognizable, before any cutting is done. Any differences in establishment may help to interpret the results that we see later on once the differences in cutting regimen start to take effect. As well as a simple list of plants per quadrat however it would be more informative to have an estimate of what percentage of the quadrat area was covered by each species.

This is rather subjective and different people would give very different estimates. A good way to be more objective is to have a grid of 25 small squares on a 1 m × 1 m frame placed on the quadrat (Fig. 4.3c). Then we record the presence or absence of each species in each small square (Fig. 4.3d). A species present in every square receives a value of 25/25 = 1, whereas one present in five squares receives 5/25 = 0.2.

Taking care in the conduct of any experiment is essential. You want to be able to tell at the end whether or not a treatment has an effect, but there are all sorts of other errors and sources of variation that could creep in. For example the treatments must be applied uniformly which is not always as easy as it sounds. The results must be recorded accurately. So if several people are involved they may need a training session and should then each record or treat one complete set of **all** treatments (otherwise the differences between treatments may be mixed up with the differences between recorders).

If you can you should try to prevent people knowing which treatment goes with which plot, for example by using plot numbers as codes. This should prevent any unconscious bias towards the treatment that someone thinks ought to come out best.

Take account of the human factor. Recording data in the field can be difficult and also boring if there is a large amount to be done. We must be prepared to carry on in strong sun or wind and even in drizzle – so it's wise to have suitable clothing and refreshments. The more comfortable you are the more likely you are to make good records. Therefore it is useful to have something soft to kneel on when peering at the plants. If you are studying the behaviour of ducks on a pond take a folding stool to sit on. If you can persuade someone to write down the observations as you call them out, this can speed the process up a great deal. Spend time designing and testing a recording sheet before photocopying it for use.

In our case the sheet needs space for the date, recorder's name and plot code, plus a column of species names down the right-hand side (Fig. 4.4). The next 25 columns are used to tick the presence of a species in one of the 25 small squares in the quadrat. The column at the far right is used to enter the total for each species at the end of the day's work. We need a minimum of 48 sheets for each time we do the recording and in practice a few spares.

Because of all the work that has gone into designing and laying out the experiment and recording the quadrats the data are precious. Ideally they should be transferred to a computer as soon as possible, whilst always keeping the original sheets. If the data cannot be input immediately the record sheets should be photocopied and the copies kept in a separate building. Always remember, what can go wrong will go wrong!

4.7 MINITAB

The observations made in the fertilizer experiment described at the beginning of the chapter can be easily compared using MINITAB. We code the three treatments (A = 1, B = 2 and C = 3) and put these codes directly into one column of the worksheet (headed 'fert' here) and the yields from each plot in the next column (headed 'yield'):

```
MTB > print c1-c2
ROW  fert  yield

  1     1    7.8
  2     1    8.3
  3     1    8.7
  4     1    7.5
  5     2    8.0
  6     2    8.9
  7     2    9.3
  8     2    9.7
  9     3   10.3
 10     3   11.8
 11     3   11.0
 12     3   10.9
```

Note that MINITAB puts in the row numbers to tell us how many observations we have. We can then ask for a boxplot to be drawn separately for each of the three treatments:

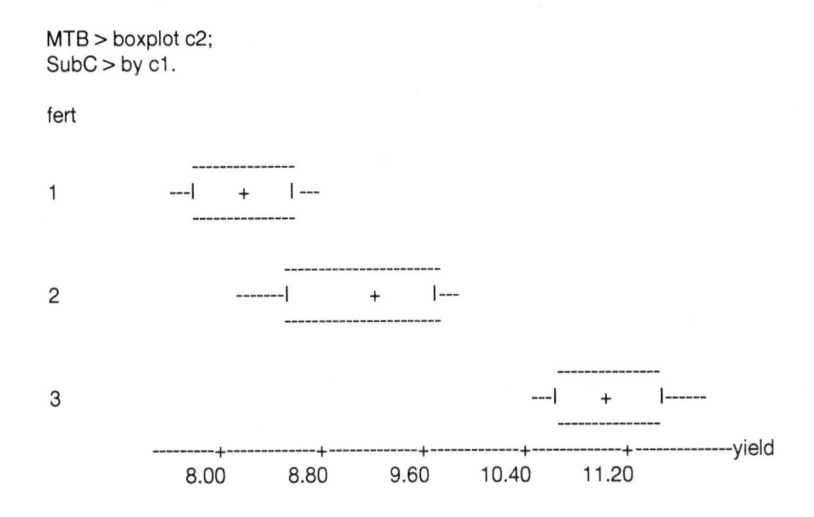

Or we could ask for the yield from each plot to be put on a graph against its treatment code:

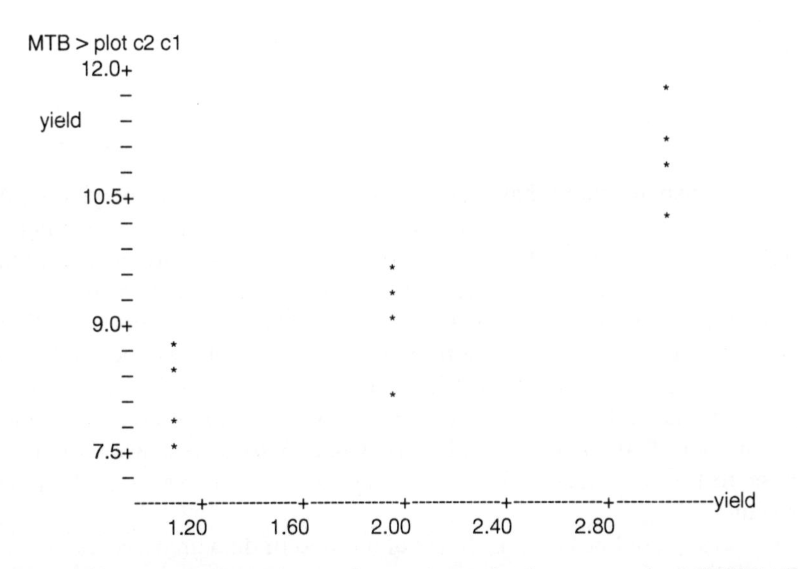

The codes for fertilizer (1, 2, 3) are along the bottom axis. MINITAB expresses the values on this axis as though it were possible to have values which were not whole numbers. This is not very elegant.

To make this clearer we can ask for the points to be labelled A, B and C.

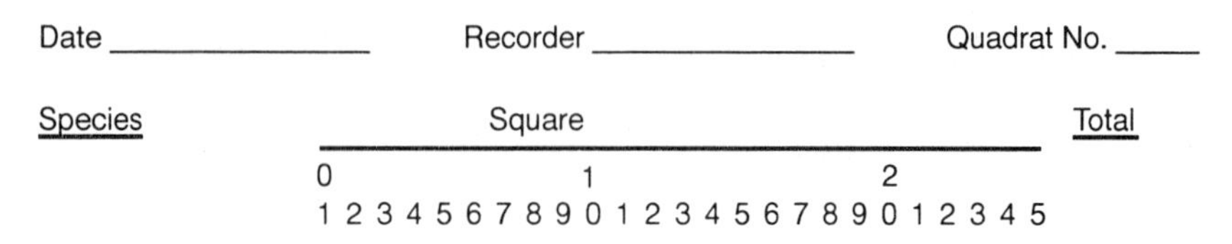

Date
Recorder
Quadrat No.
Species
Square
Total
0 1 2
1 2 3 4 5 6 7 8 9 0 1 2 3 4 5 6 7 8 9 0 1 2 3 4 5

Lolium perenne

• • •

Figure 4.4 Top of record sheet for field edge experiment.

```
MTB > l plot c2 c1 c1
        12.0+
            -                                                    C
  yield     -
            -                                                    C
            -                                                    C
        10.5+
            -                                                    C
            -
            -                         B
            -                         B
         9.0+                         B
            -   A
            -   A
            -                         B
            -   A
         7.5+   A
            -
          ---------+---------+---------+---------+---------+---------fert
                 1.20      1.60      2.00      2.40      2.80
```

These displays show that there is some overlap between the yields from treatments A and B, but that the yields from treatment C seem generally higher than those from the other two. We need some way of quantifying this description and assessing the evidence objectively. Some of the variability between yields is caused by the different amounts of fertilizer and some by random variation. How confident can we be that if we repeated this same experiment in exactly the same conditions we would obtain similar results? In Chapter 5 we will see how to describe these ideas in a model. This then will enable us (Chapter 6) to test whether or not these fertilizers affect yield and to say how certain we are about our conclusion.

General guidelines on the choice of method of data analysis are given in Appendix A.

Accounting for background variation and constructing a model

5

Experimental evidence about possible differences between treatments can only be evaluated if we first define the model we are constructing to account for the variation in our observations. In this chapter we will see how variation caused by treatments and the environment can be separated from random variation in a model. Then we will start to explain how to take account of environmental variation in an experiment. In Chapter 6 we will use the model to analyse our results.

5.1 SOURCES OF VARIATION

We design and carry out an experiment because we want to discover the important causes of variability within the system we are studying.

For example, in Chapter 4 we described an experiment to compare the effect of four treatments on the species richness of field margins.

	Cutting once/year	Cutting twice/year
With flower seeds	F1	F2
Without flower seeds	NF1	NF2

We wanted to find out whether cutting the vegetation once a year instead of twice affects the numbers of spiders which live on the site. Do they prefer vegetation which has been seeded with wild flowers to the natural grasses?

It is important to realize that even if we did not impose any treatments in the experiments we would find that the number of spiders of a particular species would not be the same in each quadrat. The natural variation across the site will mean that some plots have damper or more fertile soil or are more protected from wind and so they may have more vigorous grass growth. One quadrat may, by chance, contain many poppies because the grass had been dug up by rabbits the previous year so the poppy seeds which had been dormant in the soil for years were able to germinate at last. The number of spiders in a quadrat will be affected by these differences in the vegetation in different quadrats. Such natural variability from one quadrat or **sampling unit** to another is called **random variation**.

In a laboratory experiment such variation is often quite small. For example, the effect of different concentrations of a chemical on the growth rate of a bacterium may be studied by adding the chemical solutions to inoculated agar in petri dishes which are then kept at a constant temperature. The conditions within each petri dish will be extremely similar, apart from the deliberately imposed treatments (the chemical concentrations) being different. In contrast, in a field experiment this random variation is often very large so the quadrats and hence whatever is measured within them will differ in many ways from each other, apart from the deliberate differences introduced by the cutting and sowing treatments.

That is why if we wish to determine the factors which influence say the number of spiders per plot we need to set up a **model** of our system. This must separate background or random variation from that caused by the treatments. Let us look at the results we might collect from our field experiment. First we count the number of spiders in each quadrat in our plot. We then add together the values from each of the three quadrats per plot and divide by three to give the number of spiders per metre square. This is because it is the plots which were randomized and so they are the **experimental units**. Recording three quadrats within each plot is like sub-sampling the plots. It gives us more information and allows us to obtain a better estimate of the number of spiders per square metre of plot than if we had only recorded one quadrat. However, because each set of three quadrats are within one plot they are not independent observations and therefore should not be used separately in the **analysis of variance** that is described below. Thus we can summarize the variation in spider numbers across the 16 plots as shown in Table 5.1.

The mean number of spiders per plot in Table 5.1 is not the same in all plots. How can we make sense of the variation in terms of our model? The variation might be caused by the effects of the four treatments or it might represent random variation or, more likely, it is composed of both elements.

If we want to predict how many spiders we would find on a plot which

Table 5.1 Number of spiders per plot

	Treatment			
Replicate	F1	F2	NF1	NF2
1	21	16	18	14
2	20	16	17	13
3	19	14	15	13
4	18	14	16	12
Mean	19.5	15.0	16.5	13.0

has received a certain treatment, this experiment provides us with an estimate. It is the mean of the values from the four plots which received that treatment, the bottom row in Table 5.1.

We find that the mean number of spiders on plots which received wildflower seed and were cut once (F1) was 19.5, whereas, on those which were not sown and were cut twice a year (NF2), the mean was only 13.0. Not every plot receiving treatment F1 however will contain 19.5 spiders (this is simply an average – we cannot have half a spider in reality). There is a considerable amount of random variation around the treatment mean. Some plots have more and some less: we cannot say why they differ, except that the differences are caused by chance.

5.2 THE MODEL

We can set up a **model** which describes these ideas. If we want to predict the number of spiders on a plot the simplest estimate would be the mean of all 16 values. This is called the **grand mean**. Here it is 16.0.

We can do better than this. We observe that if a plot is in treatment F1 its mean number of spiders will be greater than the grand mean. If we calculate the difference between the treatment mean and the grand mean it is 19.5 – 16.0 = 3.5. In contrast, for treatment NF2 this difference is 13.0 – 16.0 = –3.0. We can represent this more generally as:

> Expected number of spiders in a plot of a particular treatment or the **expected value** = grand mean + (difference between treatment mean and grand mean)

This simple model predicts that each of the four plots in treatment F2 is expected to have: 16 + (15–16) = 16 + (–1) = 15 spiders per square metre. However they do not, so the model needs further work. We can find out by how much our model fails to fit our observations on each plot in treatment F2 (Table 5.2).

Table 5.2 Observed values, expected values and residuals

Replicate	Observed number	Expected number	Difference = Residual
1	16	15	1
2	16	15	1
3	14	15	–1
4	14	15	–1
Mean	15	15	0

Two of the replicate plots of this treatment (Table 5.2) have observed numbers greater than the mean for the treatment and two have values which are less. The differences between observed and expected values are called **residuals**. They represent random variation. Residuals can be positive or negative. They always add up to zero for each treatment and must also have a mean of zero.

A simple model to explain what is going on in our experiment is:

Observed number of spiders per plot = **expected number** of spiders per plot + residual.

or, in more general terms:

Observed value = **expected value + residual**

Since we already know how to calculate an expected value (see above) we can include this information as well to give the full equation:

Observed value = grand mean + (difference between treatment mean and grand mean) + residual

We can make this clearer by using the term **treatment effect** to represent the difference between the treatment mean and the grand mean.

Observed value = grand mean + treatment effect + residual

Note that both the treatment effect and residuals may be positive or negative.

Some textbooks use equations to represent the same idea:

$$Y_{ij} = \bar{y} + t_i + e_{ij}$$

The small letters i and j are called subscripts. The letter i stands for the treatment number. In our experiment we could replace it by 1, 2, 3 or 4. The letter j stands for the replicate number. We could use 1, 2, 3 or 4 here. If we wished to consider the number of spiders in treatment 2, replicate 3 we would replace these subscripts of y (the observed value) by 2 and 3 respectively. The grand mean is represented by y with a bar above it. The treatment effect is given by t with a subscript i which represents the appropriate treatment. In our case we would use 2 for treatment 2. Finally, the residual is represented by the letter e, which again has subscripts for the appropriate treatment and replicate because the random variation differs from plot to plot.

Note that the letter e is used to represent the residual because an alternative name for the residual is the **error**. This does not mean that we have made a mistake. It comes from the Latin word errare which means 'to wander'. It conveys the idea that random variation represents 'chance wandering above and below the treatment mean'.

You may also come across the term **fitted value** or **fit** instead of expected value, but they are the same thing.

The simple model we have constructed splits up the variability in our data into two parts: that which can be accounted for (expected value) and that which cannot be accounted for (resiual). Residuals can be thought of as what is left (the residue) after we have done our best to account for the variation in our data. Another way of thinking of this is to consider that we can control some sources of variation (the residuals) which are due to randomness.

5.3 BLOCKING

There are four replicates of each treatment in our experiment and the 16 plots were arranged at random with four of them on each side of the field (Fig. 4.3b). However, the four sides of the field may well provide slightly different environments. We will now see how to take this information into account in a revised layout known as a **randomized complete block design** and so reduce random variation and improve our ability to detect differences between treatments.

For example, the field may be on a slope; the side at the top may have drier or sandier soil than that at the bottom; perhaps a tall hedge runs along one side of the field, whereas there may be only a low fence along the other side. With the completely randomized distribution of treatments in Chapter 4 it is very likely that we will have allocated two replicates of a particular treatment to the field margin at the top of the slope and none to the margin at the bottom (Fig. 4.3b). (Can you calculate the chance of each treatment being present by chance only once on all four sides?) Such differences can be important because, for example, in a dry season, grass growth will be less vigorous at the top of the slope and this may mean that low-growing wild flowers have a better chance of becoming established. Thus, if our results from the treatment in which we sowed flower seeds and which we cut once (F1) show it to have a large number of herbs, this may partly be because of its over-representation on the favourable ground at the top of the slope. If we were to carry out such an experiment many times, such effects would even out since perhaps next time the same treatment might be over-represented at the bottom of the slope. However, if we have only the resources for one or two experiments we need to find a way of overcoming variations like this.

Ideally we want one replicate of each treatment to be on each of the four sides of the field. This type of design is called a **randomized complete block**. Each side of the field is a 'block' and each block contains a complete set of all the treatments within it. Each block is selected so that the conditions are even or homogeneous **within** it but that the conditions differ **between** one block and another. So in our field the top block is on slightly sandier soil and the block on one side is more shaded because of the hedge. However, these differences should not be too great. For example, one block can have a higher soil moisture than another, but if one block is a bog then the other must not be a desert: the whole experiment should be carried out on a reasonably uniform site.

How should we allocate the four treatments to the four plots within each block? This must be done using random numbers (Fig. 5.1a). This ensures that each treatment has an equal chance of occurring on each plot. We number the four plots in a block from 1 to 4. Then we use the random number button on our calculator. If the last digit is 3 we allocate treatment F1 to plot 3; if number 2 appears next we allocate treatment F2 to plot 2; number 1 would mean that treatment NF1 goes on plot 1, leaving plot 4 for treatment NF2 (Fig. 5.1b).

Such **blocking** can also be helpful in enabling us to apply treatments and to record results sensibly and fairly. Just as we have seen how to associate environmental variation with blocks, so we can do the same with our own behaviour. Perhaps it is possible to sow only four plots with flower seeds in a working day, so it will take 2 days to sow the eight plots required. We should sow two plots in each of two of the blocks on the first day (Fig. 5.2a) rather than sowing only one plot in all four blocks. Then, if it rains overnight and the soil is too wet to allow the remainder to be sown until a few days later, any differences in the establishment of plants from the two different times of sowing will be clearly associated with blocks. The same applies to recording the species present. There are 48 quadrats to be recorded. If two people are sharing the work they will have different abilities to see and to identify species correctly. One person should record

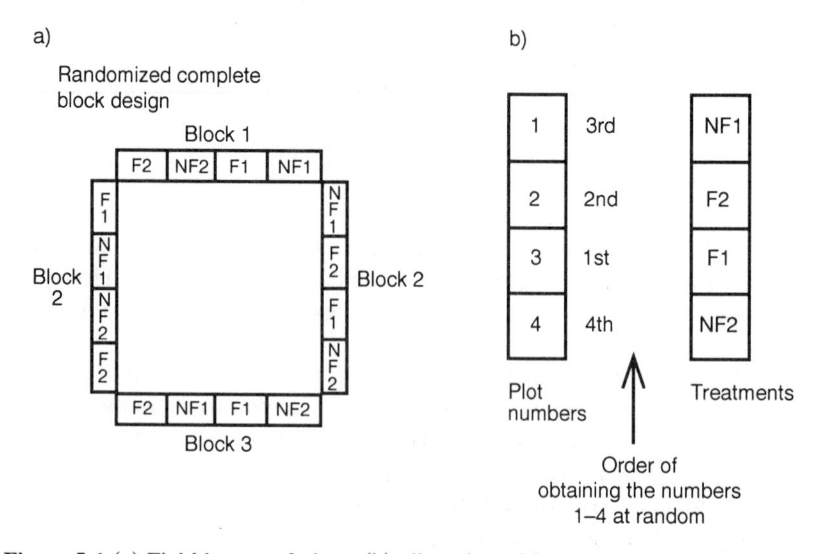

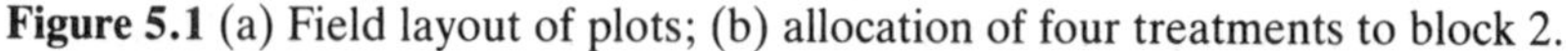

Figure 5.1 (a) Field layout of plots; (b) allocation of four treatments to block 2.

results from blocks 1 and 2 and the other from blocks 3 and 4 (Fig. 5.2b). This means that each treatment will have half of its data recorded by each person so that any differences in the recorders' abilities affect all treatments similarly on average. In addition, we can account for differences between recorders; this becomes part of the differences between blocks.

Blocking should be used wherever there may be a trend in the environment which could affect the feature in which you are interested. For example, in a glasshouse heating pipes may be at the rear of the bench, so one block should be at the rear and another at the front of the bench. Even in laboratories and growth cabinets there can be important gradients in environmental variables which make it worthwhile arranging your plots (pots, trays, petri dishes, etc.) into blocks. It is common to block feeding experiments with animals by putting the heaviest animals in one block and the lightest ones in another block. This helps to take account of differences in weight at the start of the experiment.

The concept of blocking in an experiment is similar to that of stratifying the samples taken in a sampling exercise. However, in an experiment we impose different treatments on randomly allocated plots within each block and record their effects. In a sampling exercise we simply record what is present in randomly allocated quadrats within each stratum, without imposing any treatments.

In a randomized block design we can reduce the random variation in our data by accounting for variation between blocks as well as variation between treatments. Just as plots in one treatment may tend to have more or fewer spiders than the grand mean (positive or negative treatment effect) so plots in one block may tend to have more or fewer spiders than the grand mean (positive or negative block effect). Perhaps a block at the bottom of the hill may be damper and therefore,

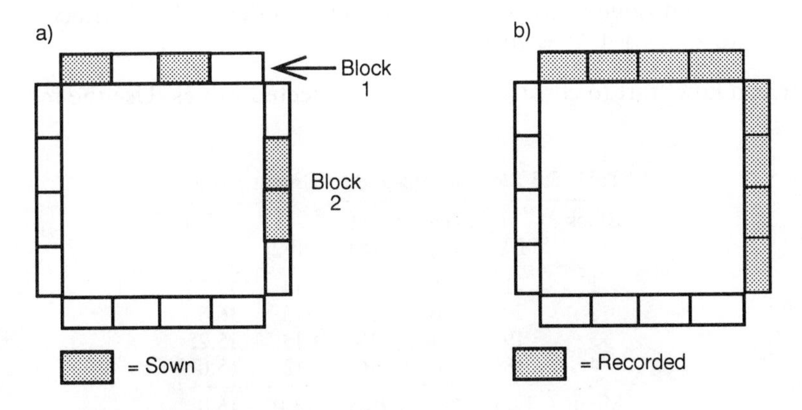

Figure 5.2 (a) Sowing – day 1; (b) recording – person 1.

whichever treatment a plot in that block receives, the vegetation grows more vigorously and tends to be more attractive to the spiders.

We now have an extra classification in our model:

Observation = grand mean + treatment effect + **block effect** + residual

$$Y_{ij} = \bar{y}\cdot + t_i + b_j + e_{ij}$$

Remember that the treatment effect = difference between treatment mean and grand mean. The block effect is defined similarly as the difference between the block mean and the grand mean.

The observations of the number of spiders per plot are again in Table 5.3. This time we have calculated both treatment and block means.

The grand mean is shown at the bottom right of the table. It is 16.0. This has not changed. We can now use our model to calculate a table, of the same layout, but showing expected values.

Expected value = grand mean + treatment effect + block effect

The block and treatment means are as we calculated them from our data. We are now predicting the expected value for each plot from our model. Let's concentrate on treatment F1 in block 1 (top left in Table 5.4). To calculate the expected value we must know the grand mean (16) and both the treatment and block effects.

Treatment effect, as before, = treatment mean – grand mean
= 19.5 – 16 = 3.5

Block effect is calculated in the same way as for the treatment effect:

Block effect = block mean – grand mean
= 17.25 – 16 = 1.25

Expected value = grand mean + treatment effect + block effect
= 16 + 3.5 + 1.25 = 20.75

We can now start to construct a table of expected values. Use the model

Table 5.3 Block means and treatment means

Block	Treatment				Mean
	F1	F2	NF1	NF2	
1	21	16	18	14	17.25
2	20	16	17	13	16.5
3	19	14	15	13	15.25
4	18	14	16	12	15.0
Mean	19.5	15.0	16.5	13.0	16.0

Table 5.4 Calculation of expected value for each plot

Block	Treatment				Mean
	F1	F2	NF1	NF2	
1	20.75	16.25			17.25
2	20.0				16.5
3					15.25
4					15.0
Mean	19.5	15.0	16.5	13.0	16.0

to calculate the expected values for treatment F2 in block 1 and for treatment F1 in block 2. You should obtain 16.25 and 20.0 respectively (Table 5.4).

There is a quicker way to calculate expected values once the first one has been calculated. It also sheds light on what the model is doing. On average all plots in treatment F2 have 4.5 fewer spiders than those in treatment F1 (19.5 – 15.0). To obtain the expected value for treatment F2 in block 1 we **subtract** 4.5 from the expected value for treatment F1 in block 1. This gives 16.25. Similarly, to obtain the expected value for treatment NF1 in block 1 we **add** 1.5 (the difference between the treatment means for F2 and NF1) to the expected value for F2 in block 1 and obtain 17.75.

The same idea works for calculating expected values in the same column. On average all plots in block 1 have 0.75 more spiders than those in block 2. Therefore the expected value for treatment F1 in block 2 is 20.75 – 0.75 = 20.0, etc. (Table 5.5).

Now we know what our model predicts. How good is it at explaining variation in spider numbers? If it was a perfect fit we would find that the tables of observed and expected values (Tables 5.1 and 5.5) were the same as each other. This is very unlikely. Usually there are differences which are the residuals. We remember that: Residual = observed value – expected value (Table 5.6).

Table 5.5 Expected values per plot

Block	Treatment				Mean
	F1	F2	NF1	NF2	
1	20.75	16.25	17.75	14.25	17.25
2	20.0	15.5	17.0	13.5	16.5
3	18.75	14.25	15.75	12.25	15.25
4	18.5	14.0	15.5	12.0	15.0
Mean	19.5	15.0	16.5	13.0	16.0

Table 5.6 Residuals per plot

Block	Treatment				Mean
	F1	F2	NF1	NF2	
1	0.25	−0.25	0.25	−0.25	0
2	0	0.5	0	−0.5	0
3	0.25	−0.25	−0.75	0.75	0
4	−0.5	0	0.5	0	0
Mean	0	0	0	0	0

The residuals in Table 5.6 seem quite small. This suggests that our model is quite good at explaining variation in spider numbers between plots; there is not much that cannot be explained either in terms of a treatment or a block effect. But how big do residuals have to be before you start to be concerned that the model is not a good fit? How do you decide whether differences between the treatments are big enough to mean anything? After all we carried out this experiment to answer the question 'Is there any difference between the effects of the four treatments on spider numbers' (because spiders are part of the wildlife that might benefit from these treatments)? In this chapter we have constructed a model to describe our experiment which seems to be relevant. This is the first step towards answering our question, but in the next chapter we need to go a step further and decide whether this model does fit the data satisfactorily and how to reach a preliminary conclusion on the significance of the apparent treatment differences.

Analysing your results – Is there anything there?

6

In Chapter 5 we saw how to construct a model to fit our experimental design. Here we will use it to analyse the data from the experiment.

We want to know whether it is reasonable to conclude that the treatments do affect spider numbers and the technique we will use is the **analysis of variance***. It was published by R.A. Fisher in 1923 and it works by splitting up the variation in all the data into pieces (components) which are attributable to (can be explained by) particular sources (for example treatments or blocks).

Analysis of variance quantifies this variation by calculating sums-of-squares. First we calculate the sum-of-squares for all the data taken together. This is called the **total sum-of-squares**: the sum of the squares of the differences between each observation and the grand mean. Do you remember these from Chapter 1? If we want to avoid the long calculation involved we can enter all 16 observations (Table 6.1) into our calculator on statistical mode; press the standard deviation button (σ_{n-1}); square it to find the variance and multiply this by n–1 (15) to find the total sum-of-squares. This reverses the procedure needed to calculate the standard deviation on p. 11. We obtain 106.0.

6.1 WHOLLY RANDOMIZED DESIGN

We will start by assuming that our data came from a wholly randomized experiment with four replicates per treatment. We will incorporate the idea of block variation later. Textbooks often refer to the analysis of a wholly randomized design as 'one-way' analysis of variance. This refers to the fact that there is only one source of variation involved (treatments).

The analysis of variance splits up the variation:

* If you have studied statistics before you may have come across a t-test. This compares the effects of two treatments. Analysis of variance is another method of doing the same thing but it can also cope with three or more treatments where it would be inappropriate to use t-test procedures.

Total sum-of-squares = treatment sum-of-squares
+ residual sum-of-squares

We have just calculated the total sum-of-squares; therefore if we can calculate the residual sum-of-squares then we can find out the treatment sum-of-squares which is what we are after because it tells us how much of the total variation is due to treatments. Therefore we need to know what the 'residuals' are. As we saw in Chapter 5 they are the differences between the observed and expected values and on p. 55 we saw how to calculate expected values. Thus the values in Table 6.1b and 6.1c can be built up.

We can now calculate the sum-of-squares of the residuals by the same method as for the total sum-of-squares. That is, we calculate the sum of the squares of the differences between the residual values in each cell of Table 6.1c and their grand mean (which is always zero). We obtain 16.0.

Table 6.1 (a) Observed values

	Treatment			
Replicate	F1	F2	NF1	NF2
1	21	16	18	14
2	20	16	17	13
3	19	14	15	13
4	18	14	16	12
Mean	19.5	15.0	16.5	13.0

(b) Expected values

	Treatment			
Replicate	F1	F2	NF1	NF2
1	19.5	15.0	16.5	13.0
2	19.5	15.0	16.5	13.0
3	19.5	15.0	16.5	13.0
4	19.5	15.0	16.5	13.0
Mean	19.5	15.0	16.5	13.0

(c) Residuals

	Treatment			
Replicate	F1	F2	NF1	NF2
1	1.5	1.0	1.5	1.0
2	0.5	1.0	0.5	0
3	−0.5	−1.0	−1.5	0
4	−1.5	−1.0	−0.5	−1.0
Mean	0	0	0	0

$$\begin{aligned}\text{Total sum-of-squares} &= \text{treatment sum-of-squares}\\ &\quad + \text{residual sum-of-squares}\\ 106.0 &= \text{treatment sum-of-squares}\\ &\quad + 16.0\end{aligned}$$

Therefore,

$$\begin{aligned}\text{Treatment sum-of-squares} &= \text{total sum-of-squares}\\ &\quad - \text{residual sum-of-squares}\\ &= 106.0 - 16.0 = 90.0\end{aligned}$$

We see that in this case the treatments account for most of the total variation. The experiment has been reasonably successful in that other sources of variation have been kept to a minimum. This won't always be the case however, so let's consider the two extreme situations we could encounter: where treatments explain **nothing** or **everything** (Fig. 6.1).

1 Treatments explain nothing

If all the treatment means are the same, knowing the treatment mean does not help you to predict the plot mean. Each treatment may contain a great deal of random variation. (Table 6.2) and the total variation would be entirely explained by this variation within treatments.

In Table 6.2 the total sum-of-squares equals the sum-of-squares of the residuals. The treatment sum-of-squares is zero. It is just as if we had selected 16 values from one population at random and allocated them as the treatment results. Note that there are many experiments where a zero treatment effect is a very desirable result; suppose for example you are testing for possible side-effects of herbicides on fish in nearby streams.

2 Treatments explain everything

At the other extreme, if treatment effects explained everything then the treatment variation would account for all the total variation and the random variation would be zero.

Table 6.2 Treatments explain nothing

	Treatment			
Replicate	F1	F2	NF1	NF2
1	16.0	15.0	16.0	17.0
2	15.0	17.0	16.0	16.0
3	17.0	16.0	17.0	15.0
4	16.0	16.0	15.0	16.0
Mean	16.0	16.0	16.0	16.0

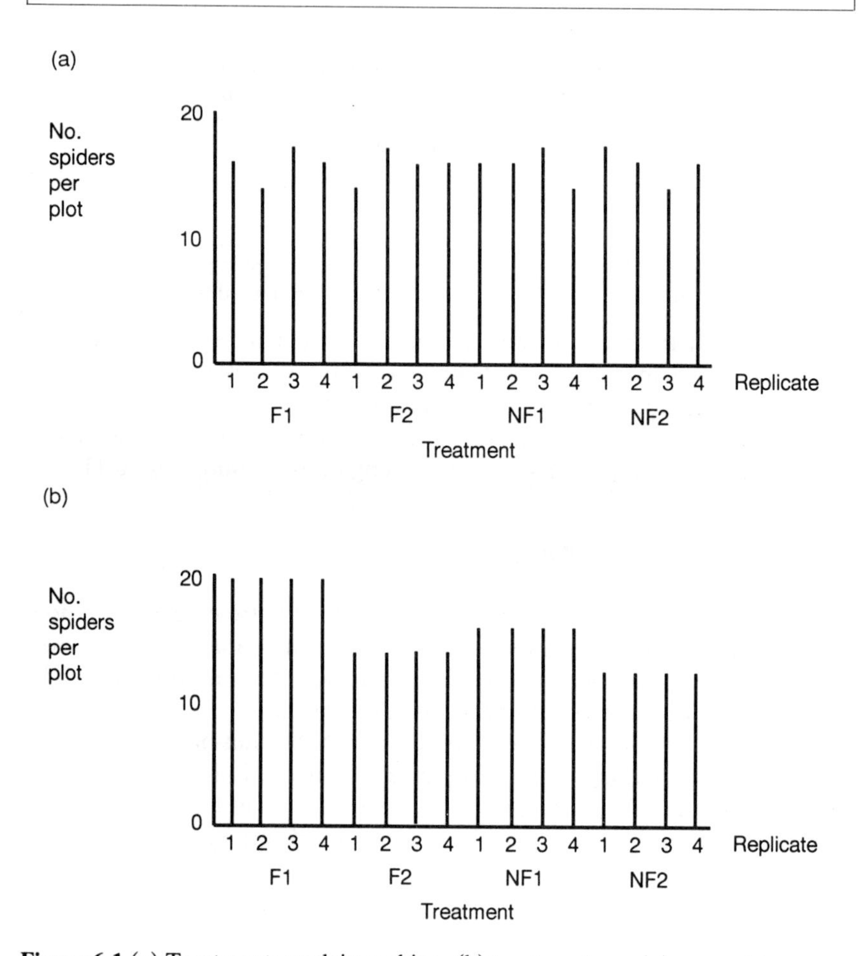

Figure 6.1 (a) Treatments explain nothing; (b) treatments explain everything.

In this situation, if you know the treatment mean you can predict the plot mean perfectly (Table 6.3). There is no random variation within any treatment. All residuals are zero because the observed and expected values are equal in each plot. The total sum-of-squares equals the treatment sum-of-squares. The sum-of-squares of the residuals (also called the residual sum-of-squares) is zero.

So far we have calculated the treatment sum-of-squares by subtracting the residual sum-of-squares from the total sum-of-squares. However, we could have calculated the treatment sum-of-squares directly. We will do this once, to show that the method works, since it can then be extended to more complicated designs (Chapter 7).

The first stage is to look at how the treatment **means** vary. Our four treatment means are: 19.5, 15.0, 16.5 and 13.0. To find the treatment

Table 6.3 Treatments explain everything

	Treatment			
Replicate	F1	F2	NF1	NF2
1	19.5	15.0	16.5	13.0
2	19.5	15.0	16.5	13.0
3	19.5	15.0	16.5	13.0
4	19.5	15.0	16.5	13.0
Mean	19.5	15.0	16.5	13.0
Total	78	60	66	52

sum-of-squares we enter the treatment **totals** into the calculator on statistical mode. These are 78, 60, 66 and 52. We then calculate their variance (using the short-cut noted earlier) by pressing the standard deviation button (σ_{n-1}) and squaring it. We then multiply by n–1 (3). This gives us the sum-of-squares of these four numbers (the treatment totals), but it ignores the fact that each of these numbers is itself the total of four observations (the four replicates). To obtain the treatment sum-of-squares on a 'per plot' scale (like the total and residual sums-of-squares), we must divide our result by 4 (the number of observations in each treatment total):

Treatment totals = 78, 60, 66, 52
standard deviation = 10.954
variance = 120
variance × 3 = 360
360/4 = 90 = treatment sum-of-squares

This is in agreement with our previous calculation for 'treatment sum-of-squares', where we obtained it by subtracting the residual sum-of-squares from the total sum-of-squares.

6.2 ANALYSIS OF VARIANCE TABLE

Whichever way we have calculated the treatment sum-of-squares we can now put it into an **analysis of variance table**. The standard method of presentation is as follows. We start by writing down the **analysis of variance plan**. This gives the structure of the experiment in terms of the sources of variation, together with their degrees of freedom:

Source of variation	df
Treatments	3
Residual	12
Total	15

Some textbooks and computer packages, including MINITAB, refer to Residual as 'Error'.

The convention is that the total variation is always on the bottom line with the residual line immediately above it. Sources of controlled variation appear above them. There are 15 degrees of freedom (df) for the total line because there are 16 plots and df = n–1. Similarly there are 3 df for treatments because there are four treatments, so n–1 = 3. Residual df are obtained by subtracting 3 df from 15 df. They are the residue. Another way of thinking about residual df is to spot that each of the four treatments has four replicates. So **within** each treatment there are n–1 = 3 df which represent random variation. So the residual variation has 3 df for each of the four treatments, giving 12 df.

It is very important to prepare an analysis of variance plan before carrying out any experiment. Otherwise, we may find out, too late, that the experiment either cannot be analysed, has too little replication to be able to detect differences between population means or has excessive replication and so is inefficient.

We can now fill in the next part of the analysis of variance table. First we put in the sums-of-squares calculated above. Then we have a column which is called **mean square**. This is the conventional name for **variance** when we are carrying out analysis of variance. We obtain it, as usual, by dividing the sum-of-squares by degrees of freedom. We only need to do this for the treatment and residual lines. So, the treatment mean square = 90.0/3 = 30.0 and the residual mean square = 16.0/12 = 1.25. The treatment and residual mean squares (or variances) can now be compared directly because both are expresses in terms of the number of independent pieces of information which they contain (n–1 = df).

If the treatment mean square was zero we would be in situation 1 above – 'treatments explain nothing'. If the residual mean square was zero, treatments would explain everything (situation 2 above). We can quantify where we are between these two extremes by dividing the treatment mean square by the residual mean square. This gives a ratio of two mean squares (or variances), a **variance ratio**. If treatments explained nothing the variance ratio = 0. If our observations had been selected from one population and allocated to treatments at random we would expect a variance ratio of 1 since treatment and residual mean squares would be equal. If treatments explained everything the variance ratio = infinity. The larger our variance ratio, the more evidence we have that the treatments differ from each other more than just by chance.

Source	df	Sum-of-squares	Mean square	Variance ratio*
Treatments	3	90.0	30.0	22.5
Residual	12	16.0	1.3333	
Total	15	106.0		

* Variance ratio = Treatments mean square/Residual mean square. Our value here is 30/1.3333 = 22.5.

We now need some way of quantifying the strength of evidence which this provides, which we do by a **hypothesis test**. It is part of the logic of statistical method that we can only assess the extent to which evidence rejects a hypothesis rather than asking to what extent our data agrees with a hypothesis. We therefore specify a hypothesis such as the following, that the four trcatmcnts comc from onc population. In other words our hypothesis is that the four treatments should have the same mean and if our samples have different means this is only due to sampling error – they differ only by chance. This is called a **null hypothesis** because it presumes no effect. It is sometimes abbreviated to H_0. The alternative hypothesis (H_1) is that the four treatments do not all come from one population. If this were true at least one of the treatments would have a different population mean from at least one of the others.

Our problem is that in carrying out an experiment, we are taking small random samples from the populations of several treatments and comparing the results. There is natural variation in all populations. Even if all the populations were identical (the null hypothesis is correct) there is a small chance that the biggest four values from a population might go into one treatment and the smallest four into another in our samples. Thus there would appear to be strong evidence from our experiment that the population means differed, when in fact they are the same. It is far more likely however that four large values from one treatment and four small values from another is evidence of a real difference between the means of the populations which we are sampling. But how much more likely?

With a variance ratio of 22.5 in the table above (p. 62) with what confidence might we reject the null hypothesis of no difference between the four treatments? If we are not using a computer program we will need to have access to a book of statistical tables or to find the appropriate table in the back of a statistics textbook. The table we want is called an ‘F table’ (Fig. 6.2). (It contains values of the statistic called ‘F’ just as a ‘t table’ contains values of ‘t’.) By comparing our calculated

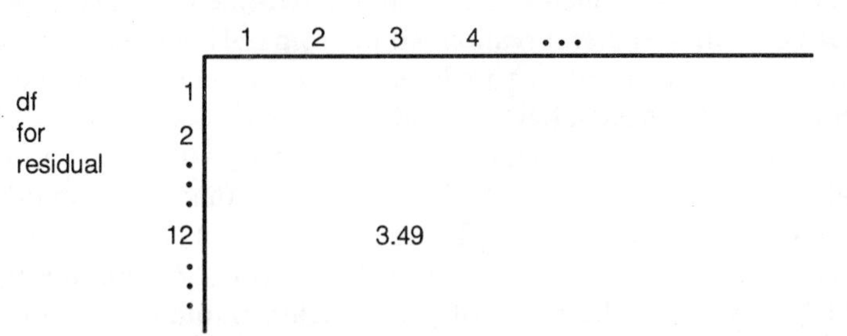

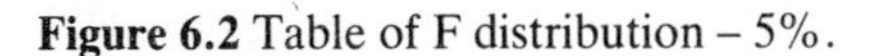
Figure 6.2 Table of F distribution – 5%.

variance ratio (22.5) with the appropriate values of F in the table we can estimate the probability 'p' of obtaining our particular results if the null hypothesis of no difference between the treatments is correct. If this probability is very small we decide to reject the null hypothesis and instead conclude that the treatments do not all come from one population.

However, what is the appropriate value (usually called the **critical value** of F) with which we should compare our variance ratio? It is an accepted standard to use the table of F labelled '5%' or 'p = 0.05' (p stands for probability; it is a proportion between 0 and 1). The numerous possible values within it depend on what combination of degrees of freedom have been used in calculating the variances of treatment and residual effects. We choose the column in the F table which has degrees of freedom for the item in which we are interested – in this case the treatment effect. Here, this is 3 df. We choose the row according to the degrees of freedom which go with the residual, in this case 12 df. This value of F – 'for 3 on 12 df' – is 3.49 (Fig. 6.2). Our variance ratio of 22.5 is much greater than this. Therefore we may conclude that we have evidence to reject the null hypothesis.

Because these tables are labelled '5%' or 'p = 0.05' we can say that if the null hypothesis were true (and the populations of the four treatments were the same) there is a less than 5% chance of our obtaining our set of results by a random selection from the populations. We are usually content to conclude that this is so unlikely that we will accept the alternative hypothesis, – H_1, and state that there is likely to be a difference between the treatment population means.

Our variance ratio is so much bigger than 3.49 that it is worthwhile checking the F table which is labelled '1%' or 'p = 0.01' or even the one labelled '0.1%' or 'p = 0.001'. These two tables give critical values of F of 5.95 and 10.8 respectively (for degrees of freedom 3 on 12). Our variance ratio is even bigger than 10.8. Thus we can conclude that there is very strong evidence to reject the null hypothesis. This is because 'p = 0.001' tells us that there is only a one in a thousand chance of our obtaining our particular set of experimental results if the treatments really all have the same effect (and so the observations come from only one population). This is such a small chance that we should prefer to conclude that the treatments do not all have the same effect on spiders.

Now we can state our degree of confidence in our conclusion in a formal way. Here we are 99.9% confident that we should reject hypothesis H_0 and accept H_1. There is only a 0.1% (p = 0.001) that there is no difference between the four populations. Computer packages like MINITAB give the value of p for each grouping in an analysis of variance from a built-in database, so we do not need to consult F tables. This means that we obtain exact probabilities. For example, 'p = 0.045' means we are less confident in rejecting the null hypothesis than if we saw '$p = 0.012$'. Often however 'p' values will be represented (particularly in published papers) by asterisks or stars. * represents $p < 0.05$, ** represents $p < 0.01$ and *** represents $p < 0.001$. However, unless you are specifically asked to present probabilities in this way you should give the exact probabilities. Note also that when results are described as 'statistically significant' this is simply a convention for describing the situation when the 'p' value is less than 0.05.

6.3 RANDOMIZED COMPLETE BLOCK DESIGNS

We can now introduce an extra line to the analysis of variance table. It will represent the amount of variation accounted for by the blocks. Textbooks often call this a 'two-way analysis of variance' because there are two classification – treatments and blocks. By accounting for variation caused by differences within the site through blocking we will reduce the random or residual variation. Therefore the residual mean square will decrease and the variance ratio for treatment effects will increase.

First, the analysis of variance plan:

Source of variation	df
Blocks	3
Treatments	3
Residual	9
Total	15

Because there are four blocks there are 3 df for blocks. We add together 3 df for blocks and 3 df for treatments and subtract the answer (6) from the total df (15) to obtain the df for the residual (9). This is 3 fewer than it was before because 3 df are now accounted for by blocks. Just as the old df for the residual has now been split into df for blocks and new df for residual so the sum-of-squares for the residual will be split: part will now belong to blocks and the remainder will be the new, smaller, residual sum-of-squares.

We must now calculate a sum-of-squares for blocks. The method is just the same as for treatments.

Our four block means are: 17.25, 16.5, 15.25 and 15.0 and the block totals are: 69, 66, 61 and 60. To find the block sum-of-squares we enter the block **totals** into the calculator on statistical mode. We then press the standard deviation button (σ_{n-1}) and square it to find the variance. We then multiply by n–1 (3). This gives us the sum-of-squares of these four totals, but, as with the treatment sum-of-squares, it ignores the fact that each of them is itself derived from four observations (one plot from each treatment). To obtain the block sum-of-squares on a 'per plot' scale (like the total, treatment and residual sums-of-squares), we must divide our result by 4 (the number of observations in each block total):

Block totals 69, 66, 61, 60
standard deviation (of block totals) = 4.243
variance = 18
variance × 3 = 54
54/4 = 13.5 = Blocks sum-of-squares

We now include the blocks sum-of-squares in the analysis of variance table. The revised residual sum-of-squares is obtained by subtracting 13.5 and 90.0 (treatment sum-of-squares) from 106.0 to give 2.5.

The blocks mean square is obtained by dividing its sum-of-squares by degrees of freedom as before. The revised residual mean square is then obtained by dividing 2.5 by 9. The revised variance ratio for treatments is obtained by dividing the treatments mean square by the revised residual mean square to give 108.0. This value is then compared with the critical value in F tables for '3 on 9 df'. Our variance ratio is very much bigger than the critical F value for $p = 0.001$. Therefore we have strong evidence for rejecting the null hypothesis. A common way of expressing this in scientific papers is to say that treatment means differ ($p < 0.001$). The < sign means 'less than'.

Source	df	Sum-of-squares	Mean square	Variance ratio
Blocks	3	13.5	4.5	16.2
Treatments	3	90.0	30.0	108.0
Residual	9	2.5	0.278	
Total	15	106.0		

The blocks mean square (4.5) though smaller than that for treatments is also very much bigger than the residual mean square. Therefore we have strong evidence that the blocks are not merely random groups. They have accounted for site variation well, in that plots in some blocks tend to contain more spiders than plots in other blocks. This source of variation has been identified and separated from the residual. Thus, the amount of random variation which remains unaccounted for is quite small.

6.4 WHICH TREATMENT IS DIFFERENT?

So far we have concentrated on **hypothesis testing** (are the treatments all the same?). This is commonly only one aspect of the problems in which we are interested and not the most important. When we carry out experiments we usually choose treatments which we are already reasonably sure will have effects which differ from each other. Formally rejecting the null hypothesis that there is no difference between the treatments is thus often a routine matter. What is frequently of greater interest is the comparison of the results for different treatments in the experiment. The treatment means are only estimates based on the samples in our experiment, so we need to calculate confidence intervals if we wish to know the range of values within which we are 95% confident that:

1. the treatment population mean lies, or
2. a difference between two population means lies.

We will explore this in Chapter 7, together with an examination of how we might ask more specific questions about treatment effects.

6.5 MINITAB PRINTOUT

The effect of four different combinations of sowing and cutting on the number of spiders.

6.5.1 The data

We type our data and code numbers for blocks and treatments (see below) into columns in the MINITAB worksheet, check and edit it and then save it onto a disc. We can then ask MINITAB to print out a copy of our information:

MTB > print C1–C3

MINITAB provides the row number on the left-hand side (column 0).

Each treatment is given a code number: 1 to 4 (column 1) and so is each block (column 2). This means that the computer program knows the correct block and treatment for each observation – in this case the number of spiders (column 3).

ROW	treat	block	spiders
1	1	1	21
2	1	2	20
3	1	3	19
4	1	4	18
5	2	1	16
6	2	2	16
7	2	3	14
8	2	4	14

9	3	1	18
10	3	2	17
11	3	3	15
12	3	4	16
13	4	1	14
14	4	2	13
15	4	3	13
16	4	4	12

6.5.2 One way analysis of variance

This assumes that the plots were not blocked and so ignores the block codes in column 2.

```
MTB > anova c3 = c1;
SUBC > means c1.
```

This asks for the data in column 3 to be explained in terms of the codes in column 1 and the means for each treatment to be printed. MINITAB responds by first summarizing the model: the factor 'treat' is 'fixed' (controlled by us) and has four levels which are coded 1, 2, 3 and 4.

Factor	Type	Levels	Values
treat	fixed	4	1 2 3 4

MINITAB assumes that column 1 contains codes for a 'fixed' or 'controlled' effect unless we tell it otherwise (see blocks below).

Analysis of variance for spiders

Source	DF	SS	MS	F	P
treat	3	90.000	30.000	22.50	0.000
Error	12	16.000	1.333		
Total	15	106.000			

MINITAB gives the heading 'F' to the variance ratio column (often labelled VR in textbooks) because the variance ratio is compared with the appropriate value in an F table. The 'p' value (0.000) is extremely small (<0.001). The chance of obtaining our sample of results if all four treatments are samples from the same population is extremely unlikely (less than one chance in 1000). We conclude that the treatments do not all have the same effect on spider numbers.

Means

treat	N	spiders
1	4	19.500
2	4	15.000
3	4	16.500
4	4	13.000

6.5.3 Two way analysis of variance

Now we introduce blocking. Thus there are now two sources of variation which we can identify: treatments and blocks. Blocks differ in nature from treatments. Treatments are fixed by us. They have a very clear meaning. In contrast, blocks are a random sample of possible plots of land in the field. We tell MINITAB this by typing the subcommand 'random C2'.

```
MTB   > anova c3 = c1 c2;
SUBC  > random c2;
SUBC  > means c1 c2.
```

Factor	Type	Levels	Values			
treat	fixed	4	1	2	3	4
block	random	4	1	2	3	4

Analysis of Variance for spiders

Source	DF	SS	MS	F	P
treat	3	90.000	30.000	108.00	0.000
block	3	13.500	4.500	16.20	0.001
Error	9	2.500	0.278		
Total	15	106.000			

Means

treat	N	spiders
1	4	19.500
2	4	15.000
3	4	16.500
4	4	13.000

block	N	spiders
1	4	17.250
2	4	16.500
3	4	15.250
4	4	15.000

The variance ratio for treatments (F = 108) is now much higher than it was in the one way analysis. We have removed variation between blocks from the random variation (error). The treatment mean square is therefore now compared with a much smaller mean square for randomness (0.278). We now have even stronger evidence for rejecting the null hypothesis of no difference between the effects of the four treatments.

6.5.4 Boxplots for treatments

As with the sampling data we can ask MINITAB to display the results for each treatment as follows:

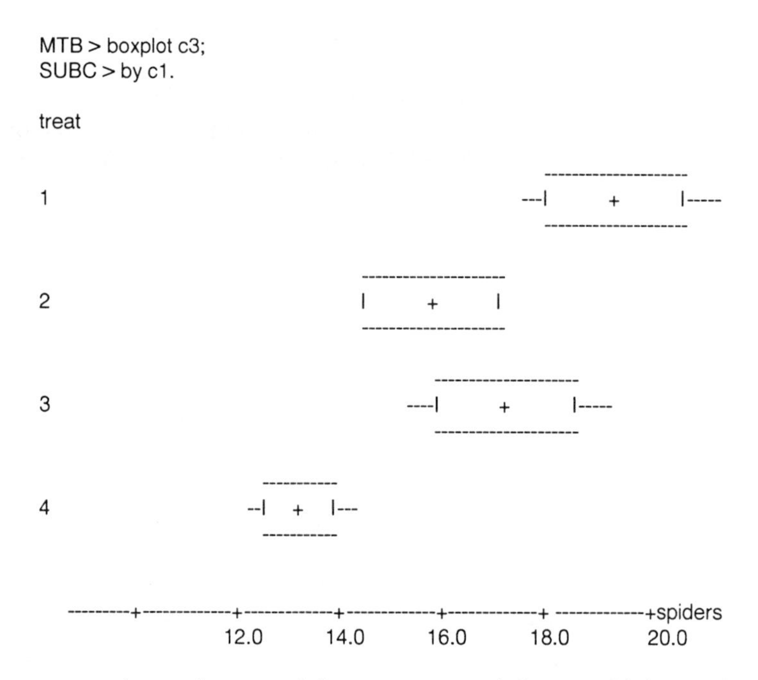

We can see how plots receiving treatment 1 have a high number of spiders while those receiving treatment 4 have a low number. While this quality of graphics in MINITAB is fine for preliminary examination of the data it can be improved for use in reports by asking for gboxplot. This produces a high resolution graphics version of the above.

We could also ask for a boxplot of the block means if these were of special interest.

1 Plotting plot values against treatments

Each plot's number of spiders can be shown against its treatment with each value represented by '*'. Where two plots in a treatment have very similar values '2' is used instead. The four treatments are coded 1, 2, 3 and 4. MINITAB presents the scale of values for treatments as a continuous one (containing decimal places). This is confusing at first sight.

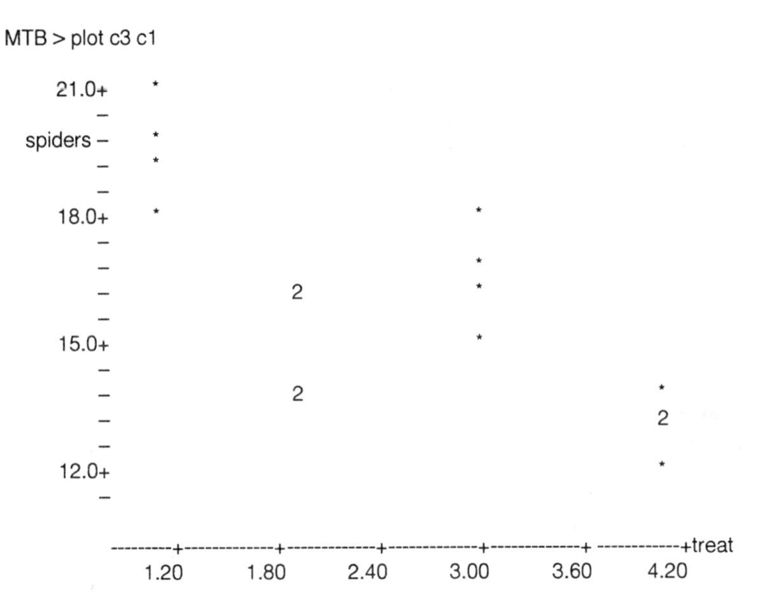

Again the use of gplot will provide a high resolution graphics version.

2 Two way analysis of variance, asking for residuals and expected (= fitted) values

When we carry out an analysis of variance we have to make certain assumptions about the observations and about the residuals. If these assumptions are untrue then the conclusions we reach are not valid. It is vital that the replicates of each treatment were independent of each other and that the experiment was randomized. We also assume that the variability within each treatment is similar. This is because we use its average (the residual mean square) to represent random variation. If this is much too small for some treatments and much too large for others we are breaking the rules. We can examine this assumption by plotting the residual for each plot against its fitted or expected value. We should see no pattern, simply a random arrangement of dots.

We now ask for MINITAB to calculate residuals and store them in column 4 and to calculate expected values and store them in column 5.

```
MTB  > anova c3 = cl c2;
SUBC > random blocks;
SUBC > resids c4;
SUBC > fits c5.
```

Factor	Type	Levels	Values
treat	fixed	4	1 2 3 4
block	random	4	1 2 3 4

Analysis of variance for spiders

Source	DF	SS	MS	F	P
treat	3	90.000	30.000	108.00	0.000
block	3	13.500	4.500	16.20	0.001
Error	9	2.500	0.278		
Total	15	106.000			

We add the name 'res' to the top of column 4 in the worksheet containing the data and the name 'fit' to the top of column 5.

MTB > print cl–c5

ROW	treat	block	spiders	res	fit
1	1	1	21	0.25	20.75
2	1	2	20	0.00	20.00
3	1	3	19	0.25	18.75
4	1	4	18	–0.50	18.50
5	2	1	16	–0.25	16.25
6	2	2	16	0.50	15.50
7	2	3	14	–0.25	14.25
8	2	4	14	0.00	14.00
9	3	1	18	0.25	17.75
10	3	2	17	0.00	17.00
11	3	3	15	–0.75	15.75
12	3	4	16	0.50	15.50
13	4	1	14	–0.25	14.25
14	4	2	13	–0.50	13.50
15	4	3	13	0.75	12.25
16	4	4	12	0.00	12.00

3 Plotting residuals against fitted values

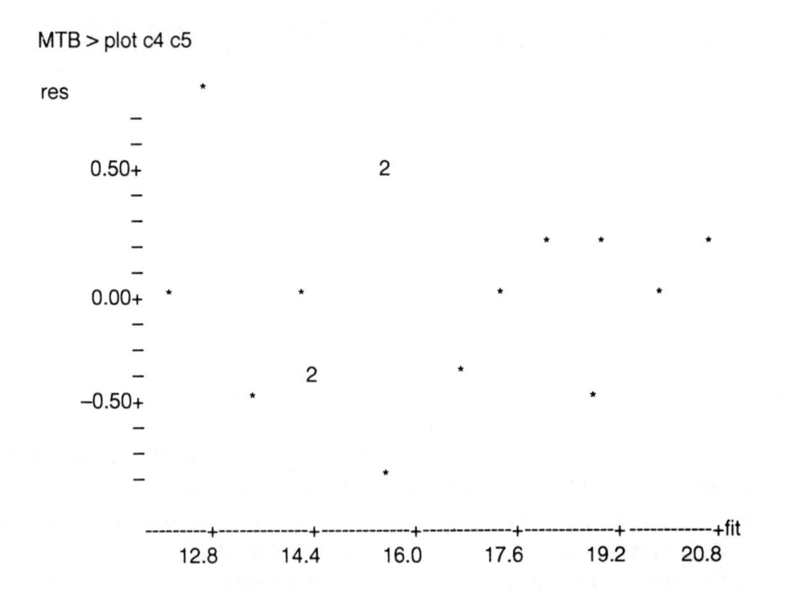

There is no obvious pattern, therefore we know that our data 'obey this rule of analysis of variance' and the conclusions we have reached from the analysis are reliable.

7 Consider the treatments of the field

In Chapter 6 we concentrated on testing the null hypothesis that 'the treatments come from the same population'. However if the results suggest that they do not, how do you decide which are the most important differences between them? In this chapter we will tackle this question by finding out how to estimate confidence intervals for population treatment means and for differences between them. Then we will see how a skilful choice of treatments allows us to obtain more specific and unambiguous information about the nature of the treatment effects.

7.1 CONFIDENCE INTERVALS

7.1.1 Treatment mean

As we saw in Chapter 2, a confidence interval is the range within which we are, for example, 95% confident that the true or population mean lies. We can calculate it by working out the estimated treatment mean from our sample and then adding and subtracting from it an amount 't × SE mean':

confidence interval = mean plus and minus (t × standard error of the mean).

The size of the confidence interval reflects the uncertainty that exists in our knowledge of the population mean because we are extrapolating from information from a few plots. If we had carried out the experiment in a slightly different part of the field, or using a different group of individual animals from the same population we would have obtained a different value for the sample mean.

Let's calculate the 95% confidence interval for a treatment where a field margin received a seed mixture and was cut once a year (treatment F1, Chapter 4). Our sample mean, from the four replicates of this treatment in the experiment, is 19.5 spiders per plot.

The value of t is looked up in t tables, for residual degrees of freedom = 12 (from the analysis of variance plan) and for the required confidence level (here the standard 95% or p = 0.05). It is 2.179.

Now we need the standard error of the mean. Unlike with a simple sample we do not find the variance by using only the four observations for a particular treatment. Rather, we use the residual mean square (= variance) from the whole experiment. This represents the pooled or average random variation within all treatments, after accounting for differences both between treatments and between blocks. It is derived from all 16 observations and therefore contains more information than if we used only the four observations for each treatment separately and so it is a better estimate of the overall variability. If we divide it by the number of replicates of a particular treatment we find the variance of the mean of that treatment. Then we take the square root to give the standard error of the mean:

Treatment mean = 19.5
t = 2.179
Residual mean square = 0.278 = variance
n = number of replicates of the treatment = 4
Variance of the treatment mean = variance/n = 0.278/4 = 0.0695
Standard error of the mean = square root of 0.0695 = 0.264

Therefore the confidence interval is:

19.5 plus or minus 2.179 × 0.264
or 19.5 plus or minus 0.575
or from 18.9 to 20.1

Thus, we are 95% confident that the population mean number of spiders per plot is in the range 18.9 to 20.1 (Fig. 7.1). There is only a 5% chance that the population mean is larger than 20.1 or smaller than 18.9.

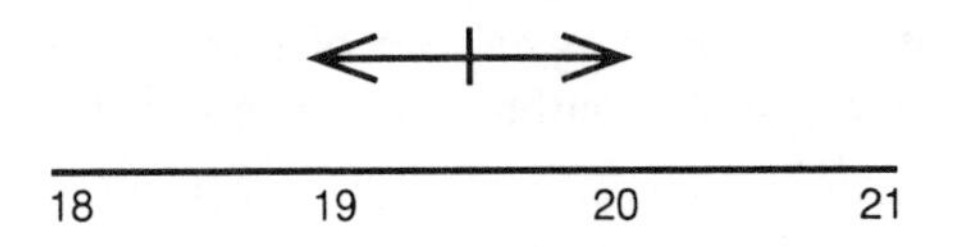

Figure 7.1 95% confidence interval for the mean number of spiders per plot on treatment F1.

7.1.2 Difference between two treatment means

Treatment NF2 (no seed mixture, cut twice per year) has an estimated mean of 13 spiders per plot. Our best estimate of the difference between the two treatments is 19.5 minus 13, which is +6.5 spiders per plot. We now wish to calculate a 95% confidence interval for the difference between the two population means:

difference plus or minus (t × standard error of the difference)

The difference is 6.5.

As usual, t is for residual degrees of freedom (12), so it is 2.179 as before.

The standard error of the difference has to take into account the variance of the first mean and the variance of the second mean, because if we are estimating how far apart two population means are and we have uncertain estimates of each of them, both amounts of uncertainty will have an effect on the range of size of the difference.

Both treatments have four replicates, therefore, in this case, the two treatment mean variances are the same as each other. We add the two variances of the means together:

$$0.0695 + 0.0695 = 0.139$$

We then take the square root, to give 0.373 as the standard error of the difference between the two means. Putting the values into the formula:

6.5 plus and minus 2.179 × 0.373
or 6.5 plus and minus 0.813
or from 5.7 to 7.3

We are therefore 95% confident that the mean difference in spider numbers between these two populations is in this range. This can be put another way: there is no more than a 5% chance that the mean difference lies outside this range and since the probabilities are equal that it lies above as below, no more than a 2.5% chance that the difference is less than 5.7. With these results, it is very unlikely that the mean difference is really zero. Thus we can be very confident that the difference between the two treatments is real and not caused by chance. Because we have 3 df we are allowed to make three particular comparisons between treatments to determine what is having the most effect on spider numbers. In this well-designed experiment there is one set of three comparisons which is most informative. We will now outline the reasons for this.

7.2 FACTORIAL STRUCTURE OF TREATMENTS

We can think of a treatment as the experience which an animal or plant (the experimental unit) receives from us. Some treatments may be the result of several 'things' being applied. Thus we now need to introduce a new technical term – **factor**. For example, in our experiment we have four treatments. However these are in fact the four combinations resulting from the application of two **FACTORS** (SOWING and CUTTING) which are each present at different **levels**. (Levels refers to how a factor is applied. In this case sowing is applied at two levels: sown or unsown, and cutting is similarly applied at two levels: cut once or cut twice (Table 7.1).) In other experiments more levels may be involved. For example a drug may be given at four levels (0, 2, 4 or 8 mg per patient) or animals might receive one of three levels of feed supplement (0, 6 or 12 g per day). The zero level is as important as any other level.

Table 7.1 Factorial structure of treatments

Factor 2 CUTTING	Factor 1 SOWING	
	Level 1 Sown	Level 2 Unsown
Level 1 Cut once	Treatment 1 F1	Treatment 2 NF1
Level 2 Cut twice	Treatment 3 NF1	Treatment 4 NF2

Returning to our vegetation management experiment, this has a 2 × 2 factorial structure (because each of the factors is present at two levels). Since we have four replicates of each treatment the experiment has 16 observations (2 × 2 × 4).

To summarize:

FACTOR = something which may have an effect (for example, herbicide, moisture)
LEVEL = state of a factor (for example, present or absent; low, medium or high)
TREATMENT = a particular combination of one level of one factor plus one level of another factor.

It will help to remember that, when there is a factorial structure, the word treatment can be replaced by the phrase 'treatment combination'.

Another example of a factorial structure is shown in Table 7.2. In this case the number of levels is different for the two factors being three for fibre and four for protein.

Table 7.2 3 × 4 factorial, with twelve treatment combinations

		Fibre in diet	
Protein in diet	low	medium	high
low	5 reps	5 reps	5 reps
medium	5 reps	5 reps	5 reps
high	5 reps	5 reps	5 reps
very high	5 reps	5 reps	5 reps

Each treatment combination then has five replicates so the structure for this experiment can be summarized as a '3 × 4 factorial' with five replicates.

Our vegetation management experiment has 3 df for treatments (one less than the number of treatment combinations). If the variance ratio for treatments is significant we have evidence to show that at least one of the four treatments has a population mean which is different from the population mean of at least one other treatment. This is ambiguous. Immediately we want to ask more specific questions. We can ask three independent questions because we have 3 df. Thus if we take the treatment line from the analysis of variance table on p. 61 it can be broken down as follows:

Source	df
Sown *versus* unsown	1
Cut once *versus* cut twice	1
Interaction sow x cut	1
4 Treatments	3

Each line represents a separate question which we can ask. First, in general, is there evidence that the sown plots have a different number of spiders per plot than the unsown ones? This component of the treatment sum-of-squares represents the **main effect** of sowing averaged over the two types of cutting. It ignores the fact that four of the eight sown plots were cut once while the other four were cut twice, and similarly for the unsown plots. It simply asks, in general, did sowing a wildflower mixture have any effect?

The second line represents the main effect of cutting. In other words, we compare the spider numbers on the eight plots which were cut once with those on the eight plots which were cut twice. Does cutting once, in general, have a different effect for cutting twice?

The third line represents a very important idea. It enables us to see whether cutting once instead of twice changes spider numbers similarly on all plots irrespective of whether they were sown with wildflower seed or not. If this test is not significant we can assume that the two factors are **independent**. If this test is significant we say that the two factors **interact**; in other words the effect of cutting is not consistent, rather it depends on whether seed has been sown or not. For example, suppose that spiders liked to hunt on the wild flowers (because flies are attracted to the flowers) but the second cut in the plots that receive it removes the flower heads – hence no flies, hence no great abundance of spiders. In this case an interaction would be expected – sowing wild flowers would give more spiders than on unsown plots when plots were also cut only once but this would not be so in the plots which were cut twice.

7.2.1 How to split up treatment sum-of-squares to assess main effects and interaction

We could calculate the sums-of-squares for the sowing main effect using the same method as for treatments. In this case we use two totals: that for the eight plots cut once (36) and that for the eight plots cut twice (28). We enter these into the calculator; square their standard deviation; multiply the result by one (degrees of freedom for sowing) and divide it by eight (as there are eight observations in each total). This gives a value of 64 for the cutting sum-of-squares. Following the same process for the totals for the eight unsown and eight sown plots gives a sowing sum-of-squares of 25.

Inserting these into the analysis of variance table (below) shows us that the interaction sum-of-squares can then be calculated by subtracting both the sowing and cutting sum-of-squares from the treatment sum-of-squares:

Treatments sum-of-squares = sowing sum-of-squares + cutting sum-of-squares + sowing × cutting interaction sum-of-squares.

Interaction sum-of-squares = Treatments sum-of-squares minus sowing sum-of-squares minus cutting sum-of-squares.

Interaction sum-of-squares = 90 – 25 – 64 = 1.

Source	df	Sum-of-squares
Sown *versus* unsown	1	25
Cut once *versus* cut twice	1	64
Interaction sow x cut	1	1
4 Treatments	3	90
Residual	12	16
Total	15	106

We find the mean square for each of the three component lines of the treatment line by dividing their sums-of-squares by their degrees of freedom. As in this case df are always 1 the mean squares equal the sums-of-squares. If, as with the structure in Table 7.2, factors have more than two levels then the df will not be 1 and hence mean squares will not equal sums-of-squares. For example in Table 7.2 the main effect for fibre in the diet has 2 df as there are three levels.

Then each mean square is divided by the residual mean square from the original analysis (p. 62) to produce three variance ratios which can each be compared with the critical F value from Tables for 1 on 9 df (5.12). The annotated first MINITAB printout at the end of this chapter shows how the computer package carries out all these calculations for us.

7.2.2 Why bother with a factorial structure for your experiment?

A factorial structure is a very efficient way of investigating biological (and other) phenomena. The cunning trick is that there is ‘hidden replication’.

For example, although there are only four replicates of each of the four combinations of sowing and cutting, there are eight replicates for 'with seed' and eight replicates for 'without seed' in the above experiment. Also, we save time and money by carrying out one experiment in which two factors are explored, instead of examining each separately. Above all we have the opportunity to find out whether these factors act independently of each other or whether they interact. If the former is true, life is simple and we can make general recommendations like 'we should sow flower seed to increase spider numbers'. If the latter is true it is important to discuss main effects only in the light of their interactions – we cannot generalize about sowing seed; its effect on spider numbers depends on how frequently we cut the vegetation. A complete factorial structure can have many factors, each at many levels but they must be balanced: all possible combinations of factors at all levels must be present. If any are missing then the analysis described above will not work. However, MINITAB can be used to analyse such unbalanced data whether the lack of balance is intentional or accidental (as when one or more plots are lost because of an accident). An example of such an analysis is given at the end of this chapter.

A useful check on the degrees of freedom in the ANOVA plan is that the number of df for an interaction can be found by multiplying together the number of df for each of the component factors (hence above $1 \times 1 = 1$ or in Table 7.2 : $2 \times 3 = 6$). A more obvious way of obtaining the number of interaction df is to take away the df for all main effects and all other interactions from the treatment df. Therefore in Table 7.2: $11 - 2 - 3 = 6$.

7.3 MINITAB PRINTOUTS

7.3.1 The effect of cutting and sowing on spider numbers

1 Factorial analysis of variance, using codes for sowing and cutting levels

MTB > print c1–c7

ROW	treat	block	spiders	res	fit	seeds	cut
1	1	1	21	0.25	20.75	1	1
2	1	2	20	0.00	20.00	1	1
3	1	3	19	0.25	18.75	1	1
4	1	4	18	–0.50	18.50	1	1
5	2	1	16	–0.25	16.25	1	2
6	2	2	16	0.50	15.50	1	2
7	2	3	14	–0.25	14.25	1	2
8	2	4	14	0.00	14.00	1	2
9	3	1	18	0.25	17.75	0	1
10	3	2	17	0.00	17.00	0	1
11	3	3	15	–0.75	15.75	0	1

12	3	4	16	0.50	15.50	0	1
13	4	1	14	–0.25	14.25	0	2
14	4	2	13	–0.50	13.50	0	2
15	4	3	13	0.75	12.25	0	2
16	4	4	12	0.00	12.00	0	2

```
MTB >anova c3 = c6  c7;
SUBC > resids c4;
SUBC > fits c5;
SUBC > means c6  c7.
```

Column 6 is seeds and column 7 is cut. The symbol between c6 and c7 is called the pipe symbol. Its meaning here is: please examine the main effect of c6, the main effect of c7 and the interaction between them. Residuals are put into column 4 and fitted values into column 5. These have been named by us in the worksheet.

Factor	Type	Levels	Values	
seeds	fixed	2	0	1
cut	fixed	2	1	2

Analysis of Variance for spiders

Source	DF	SS	MS	F	P
seeds	1	25.000	25.000	18.75	0.001
cut	1	64.000	64.000	48.00	0.000
seeds* cut	1	1.000	1.000	0.75	0.403
Error	12	16.000	1.333		
Total	15	106.000			

The p values for seeds and for cut are both very small, showing that there is very strong evidence that both factors affect spider numbers. However, the p value for the interaction is much greater that 0.05, showing that there is no evidence for an interaction between the factors. The main effects may be interpreted independently of one another.

MEANS

seeds	N	spiders
0	8	14.750
1	8	17.250

cut	N	spiders
1	8	18.000
2	8	14.000

seeds	cut	N	spiders
0	1	4	16.500
0	2	4	13.000
1	1	4	19.500
1	2	4	15.000

We can plot the number of spiders in each plot against the two types of sowing.

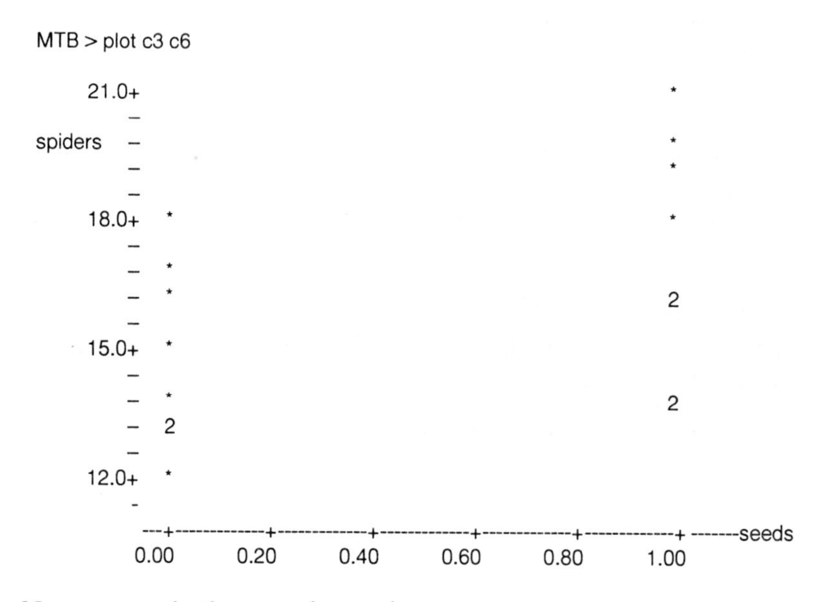

Now we can do the same for cutting.

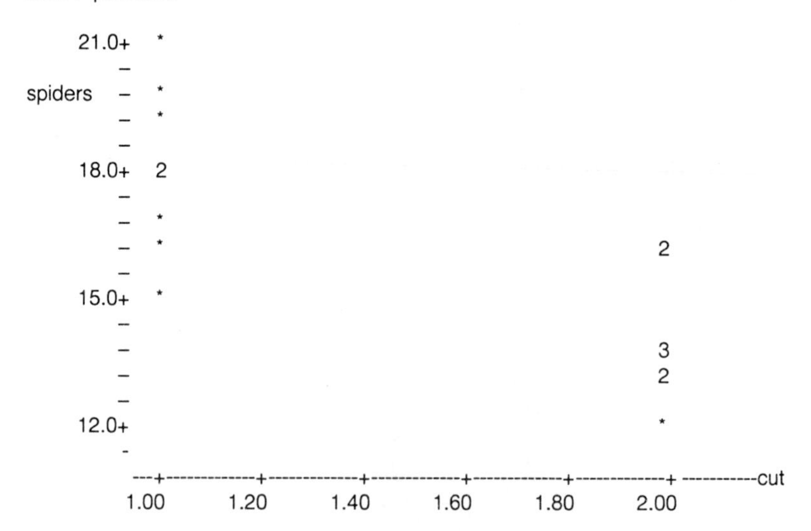

7.3.2 The effect of amount of phosphate and contact with plant tissue on bacterial growth

We have designed an experiment to investigate the importance of increasing amounts of phosphate (1, 2, 3 and 4 mg per culture) and of contact with living plant tissue (present or absent) on the extension of a bacterial colony (mm per day).

The experiment was carried out by inoculating a carbohydrate medium on petri dishes with the bacterium. One set of each of the eight treatments was established on each of 3 days, making 24 dishes in all. The amounts of growth are analysed in MINITAB:

```
MTB > print c1-c4

ROW   expt   contact   phos   exten   treat

  1     1       1        1      10      1
  2     1       0        1       6      2
  3     1       1        2      13      3
  4     1       0        2      11      4
  5     1       1        3      14      5
  6     1       0        3      20      6
  7     1       1        4      16      7
  8     1       0        4      22      8
  9     2       1        1      12      1
 10     2       0        1      10      2
 11     2       1        2      13      3
 12     2       0        2      13      4
 13     2       1        3      14      5
 14     2       0        3      14      6
 15     2       1        4      14      7
 16     2       0        4      18      8
 17     3       1        1      14      1
 18     3       0        1      10      2
 19     3       1        2      12      3
 20     3       0        2      10      4
 21     3       1        3      10      5
 22     3       0        3      14      6
 23     3       1        4      16      7
 24     3       0        4      18      8
```

```
MTB > anova c4 = c1 c2  c3;
SUBC > random c1;
SUBC > resids c5;
SUBC > fits c6;
SUBC > means c1 c2  c3.
```

Factor	Type	Levels	Values			
expt	random	3	1	2	3	
contact	fixed	2	0	1		
phos	fixed	4	1	2	3	4

Analysis of Variance for exten

Source	DF	SS	MS	F	P
expt	2	4.000	2.000	0.41	0.670
contact	1	2.667	2.667	0.55	0.471
phos	3	166.000	55.333	11.39	0.000
contact*phos	3	57.333	19.111	3.93	0.031
Error	14	68.000	4.857		
Total	23	298.000			

While there is very strong evidence for phosphate having an effect ($p<0.001$) in general this is modified by the presence or absence of plant material (interaction $p = 0.031$).

MEANS

expt	N	exten
1	8	14.000
2	8	13.500
3	8	13.000

contact	N	exten
0	12	13.833
1	12	13.167

phos	N	exten
1	6	10.333
2	6	12.000
3	6	14.333
4	6	17.333

contact	phos	N	exten
0	1	3	8.667
0	2	3	11.333
0	3	3	16.000
0	4	3	19.333
1	1	3	12.000
1	2	3	12.667
1	3	3	12.667
1	4	3	15.333

We need to plot the eight treatment means to see the interaction effect (code numbers 1 to 8 in column 5).

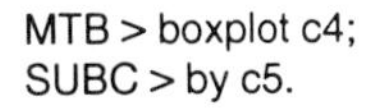

```
MTB > boxplot c4;
SUBC > by c5.
```

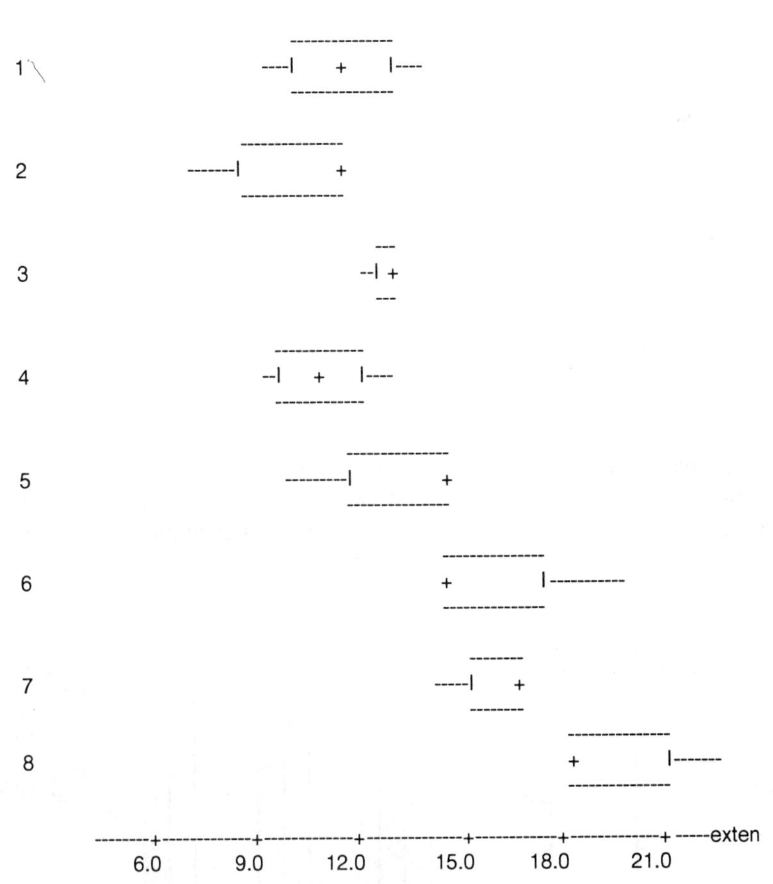

The bacteria not in contact with plant tissue (even numbered treatments: 2, 4, 6 and 8) show a great response to increasing amounts of phosphate. In contrast, the bacteria which are in contact with plant tissue (odd-numbered ones: 1, 3, 5 and 7) show much less response. Perhaps this is better displayed in an *interaction diagram* (Fig. 7.2). We might speculate on the biological reasons for such a difference.

MTB > plot c6 c7

We can plot residuals against fitted values. If, as here, there is no pattern, this reassures us that the assumption of the treatments having similar variances is a reasonable one.

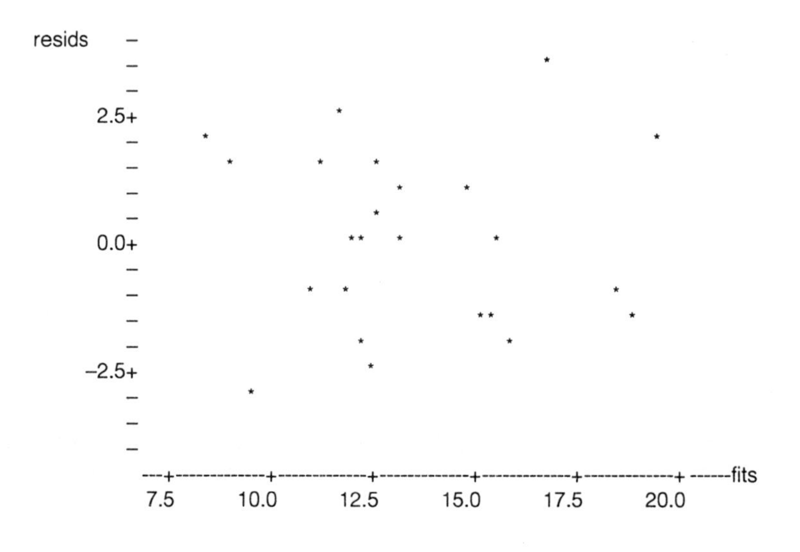

Other particular comparisons between treatment means or groups of treatment means can be made but this should be done with care. Ask a statistician for advice before carrying out your experiment. He or she will help you answer your questions efficiently and validly.

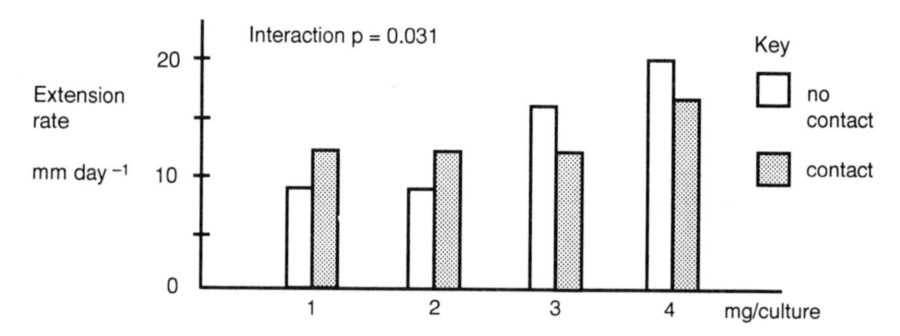

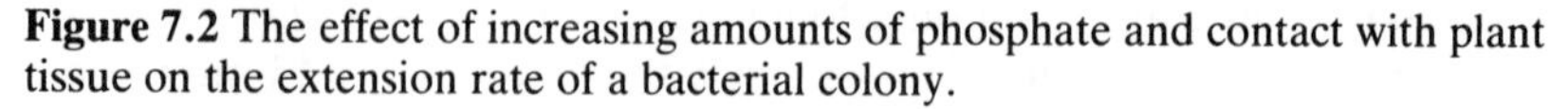
Figure 7.2 The effect of increasing amounts of phosphate and contact with plant tissue on the extension rate of a bacterial colony.

7.3.3 What to do if you have missing observations

1 Observations of zero

It is important to realize the difference between missing values caused by accidents and values of zero in a dataset. The latter may represent death of an individual and this may be related to the treatment. If so, the 0 g is as valid an observation as 5.3 g or 10 g and should not be omitted. Problems can arise when there are quite a few zero values in a dataset. If these are clustered in one or two treatments it is best to exclude those treatments from the analysis – since they are obviously different from the rest.

2 *Mistakes and disasters*

These will happen. Despite taking care with your experiment, something may well go wrong. You accidentally drop a test tube on the floor, the tractor driver accidentally ploughs the wrong plot or your helper throws away a paper bag containing the shoots from one plant before you have weighed them. You now have the problem of one (or more) **missing observations**. It is sensible to represent these by a '*' on your record sheet (with a footnote describing the problem) so that you do not confuse them with a real value of zero.

Since you have fewer observations you have less information and so the experiment is less likely to be able to detect differences between the populations which you are comparing. Another problem is that you now have more information about some treatments than about others. A one-way analysis of variance can be carried out in MINITAB as usual. It takes into account the differing replication between the treatments.

In a wholly randomized experiment we compared the effects of six different plant hormones on floral development in peas. A standard amount was applied to the first true leaf of each of four plants for each hormone, making 24 plants in the experiment. Unfortunately, two plants were knocked off the greenhouse bench and severely damaged: one from hormone 2 and one from hormone 6. So we have data on the number of the node (the name given to places on the stem where leaves or flowers may appear) at which flowering first occurred for only 22 plants. This is in C3 of a MINITAB worksheet, with treatment codes in C2.

```
MTB > print c2 c3

ROW   hormone   node
  1      1       49
  2      2       53
  3      3       49
  4      4       53
  5      5       51
  6      6       58
  7      1       52
  8      3       50
  9      4       57
```

10	5	56
11	1	51
12	2	54
13	3	52
14	4	54
15	5	51
16	6	54
17	1	48
18	2	57
19	3	49
20	4	51
21	5	54
22	6	55

MTB > anova c3 = c2;
SUBC > means c2.

Factor	Type	Levels	Values
hormone	fixed	6	1 2 3 4 5 6

Analysis of variance for node

Source	DF	SS	MS	F	P
hormone	5	101.008	20.202	4.61	0.009
Error	16	70.083	4.380		
Total	21	171.091			

MEANS

hormone	N	node
1	4	50.000
2	3	54.667
3	4	50.000
4	4	53.750
5	4	53.000
6	3	55.667

The analysis of variance provides strong evidence to reject the null hypothesis that all six hormones affect floral initiation similarly. If we wish to calculate standard errors for treatment means we must remember that they are based on different numbers of replicates. Remember the formula is:

$$SE\ mean = \sqrt{\frac{residual\ mean\ square}{n}}$$

For hormones 2 and 6, where we have only three replicates, the SE for each mean is the square root of 4.38/3 = 1.46. Whereas, for the remaining hormones it is the square root of 4.38/4 = 1.095, which is considerably smaller.

Perhaps we realized that our glasshouse was not a homogeneous environment and so laid out our experiment in four randomized complete blocks. We now have a two way analysis of variance. However, this is not **balanced**:

	Hormone					
Block	1	2	3	4	5	6
1	49	53	49	53	51	58
2	52	*	50	57	56	*
3	51	54	52	54	51	54
4	48	57	49	51	54	55

If we want to compare the mean of hormone 2 with that of hormone 4 we have a problem. It may differ because the hormone has a different effect but it may also differ because hormone 2 was not represented in block 2 which could be a location in the glasshouse which is especially good or especially bad for plant growth. Here are the data with block and treatment codes in MINITAB:

```
MTB > print c1–c3
```

ROW	block	hormone	node
1	1	1	49
2	1	2	53
3	1	3	49
4	1	4	53
5	1	5	51
6	1	6	58
7	2	1	52
8	2	3	50
9	2	4	57
10	2	5	56
11	3	1	51
12	3	2	54
13	3	3	52
14	3	4	54
15	3	5	51
16	3	6	54
17	4	1	48
18	4	2	57
19	4	3	49
20	4	4	51
21	4	5	54
22	4	6	55

We can display the data:

MTB > boxplot c3;
SUBC > by c2.

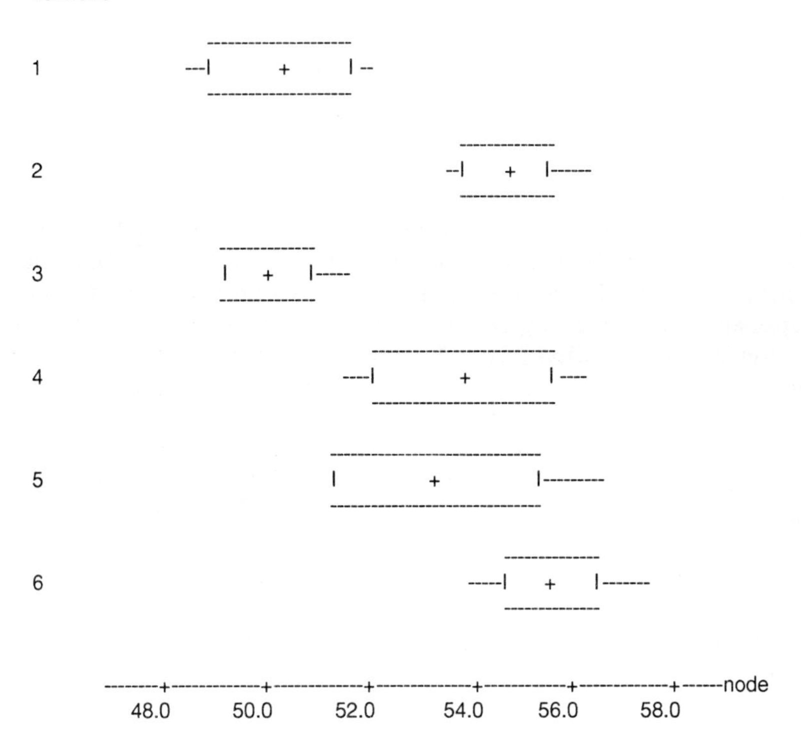

If we try to analyse the data using the usual anova command in MINITAB:

MTB > anova c3 = c2 cl

we receive a surprise

* ERROR * Unequal cell counts

This is because there isn't one representative of each treatment in each block. To analyse these data we need to use a more flexible program. MINITAB has a command 'glm' which stands for general linear model. If this is used in place of anova, all will be well:

```
MTB  > glm c3 = c2 cl;
SUBC > resids c4;
SUBC > fits c5;
SUBC > means cl c2.
```

Factor	Levels	Values
hormone	6	1 2 3 4 5 6
block	4	1 2 3 4

Analysis of Variance for node

Source	DF	Seq SS	Adj SS	Adj MS	F	P
hormone	5	101.008	117.632	23.526	6.56	0.003
block	3	23.465	23.465	7.822	2.18	0.139
Error	13	46.618	46.618	3.586		
Total	21	171.091				

The puzzling thing about this response is that there are two columns of sums-of-squares. These are 'sequential' (Seq) and 'adjusted' (Adj). When an experiment is balanced the order in which we ask the terms in the model (here hormones and blocks) to be fitted doesn't affect the outcome. There is only one value for the sum-of-squares due to hormones, whether we ask it to be fitted either before or after blocks. However, now we have an unbalanced experiment the order in which the terms are fitted is very important.

In the sequential sum-of-squares column hormone has been fitted first and then blocks. This was the order in which we asked for the two terms in the glm line. However, in the adjusted sum-of-squares column each term has the sum-of-squares appropriate to it if it were fitted last in the model. Here, if hormones is fitted before blocks its SS is 101.0, but if it is fitted after blocks its SS is 117.6. It is sensible to use the adjusted sum-of-squares here. This is appropriate since it represents evidence for differences between the hormone effects after taking into account the fact that some hormones were not represented in all of the blocks.

Let's now look at the hormone means. Although the data are identical with the figures used in the one way analysis, the means are not the same:

Means for node

block	Mean	Stdev
1	52.17	0.7731
2	55.14	0.9981
3	52.67	0.7731
4	52.33	0.7731
hormone		
1	50.00	0.9468
2	55.35	1.1270
3	50.00	0.9468
4	53.75	0.9468
5	53.00	0.9468
6	56.35	1.1270

For example, the mean for hormone 2 has increased from 54.667 to 55.35. This adjustment takes account of the fact that the representative of hormone 2 in block 2 was accidentally lost. Block 2 was, in general, one in which, for those treatments which were present, flowering was initiated at a higher node number. It is only fair that we should adjust the value for hormone 2 upwards. This method provides us with an estimate of the results we might have expected to obtain if all hormones were present in all blocks.

We plot residuals against fitted values:

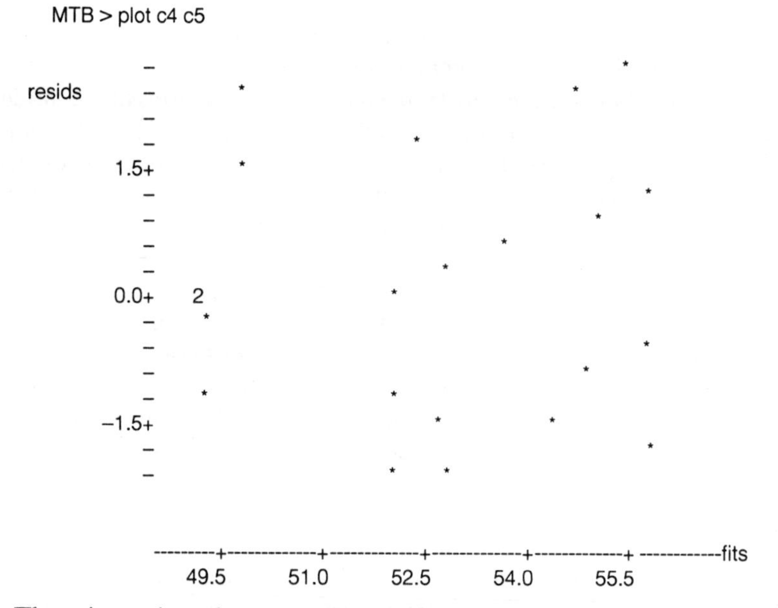

There is no sign of pattern. We ask for a histogram of the residuals:

MTB > hist c4

Histogram of resids N = 22

Midpoint	Count	
–2.0	3	***
–1.5	3	***
–1.0	2	**
–0.5	2	**
0.0	4	****
0.5	1	*
1.0	2	**
1.5	2	**
2.0	0	
2.5	3	***

These are not showing a normal distribution; there are rather too many very big and very small values. However, with only 22 values we would not be too concerned about such slight imperfections.

8 Relating one thing to another

The previous chapters have been concerned with estimating the effects of different treatments (for example frequency of cutting vegetation) and testing the null hypothesis that they all have the same effect (for example on the number of spiders). Here we will see what to do when our treatments consist of increasing amounts or 'levels' of a factor (for example water on plants, doses of a drug given to humans).

In this case we want to describe the nature of the relationship (if any) between the amount of water and plant growth or between the dose of the drug and the time taken for symptoms of disease to disappear.

The first thing to do is to plot our observations. The usual practice is to put the **response variable** (plant weight or time for symptoms to disappear) on the left-hand or y axis (also called the ordinate) and the amount (of water or drug) on the horizontal or x axis (also called the abscissa). The response variable is also known as the dependent variable because it may depend on (be affected by) the amount of the substance applied (the independent variable).

If there is no relationship between the two variables the plot of response against amount will show a random scatter of points (Fig. 8.1a). The points should not form any pattern. If we find this, we do not need to carry out any further analysis. However, we may see a pattern. The simplest pattern is where the points seem to lie about a straight line. If this line is horizontal it means that knowing the amount of substance applied does not help us to predict the size of the response (Fig. 8.1b). However, the line may slope up (a **positive** relationship in which more of x tends to produce more of y) or down (a negative relationship in which more of x leads to less of y) (Figs 8.1c and d). We then need to find a way of describing the relationship so that we can make use of it in future to predict just how much more (or less) y we will get for a given amount of x.

Another possibility is that the points seem to lie on a curve. Perhaps plant growth is increased by giving them more water, but only up to a certain amount (Fig. 8.2e) of water, above which growth cannot increase further. Similarly, the percentage body fat in a person may become less

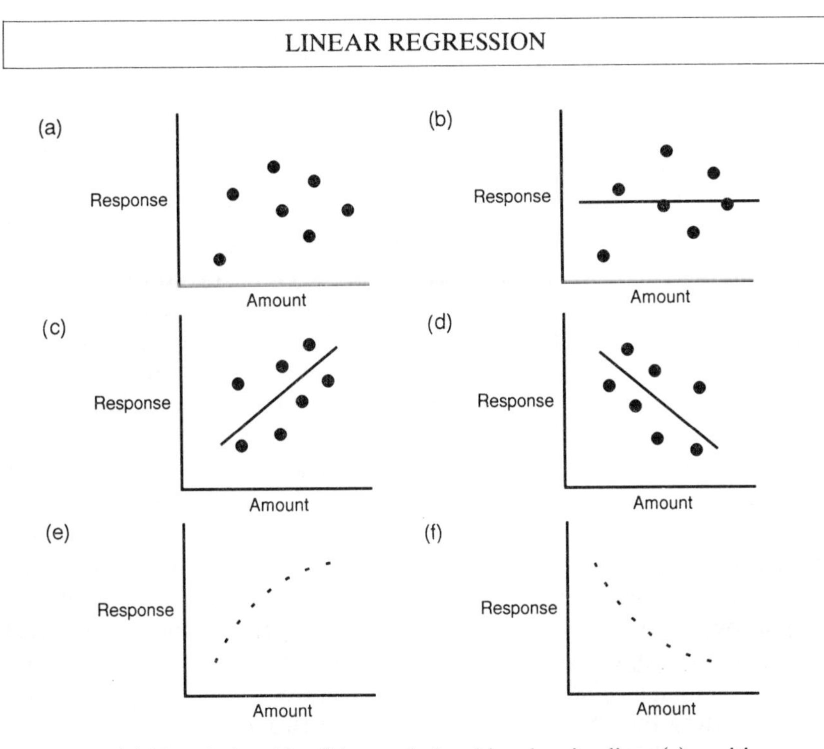

Figure 8.1 (a) No relationship; (b) no relationship, showing line; (c) positive relationship, with line; (d) negative relationship, with line; (e) response increases, then levels off; (f) response decreases, then levels off.

with increasing exercise but there is a minimum percentage fat which cannot be reduced (Fig. 8.1f).

First we will consider how to test whether our results provide evidence for a significant straight-line relationship (one which will help us to explain or predict the response by knowing the amount of the factor applied) or not. Remember we have only obtained observations from a few individuals but we wish to be able to extrapolate to the whole population. The name given to this process of providing the '**best-fit straight line**' which fits the points is **linear regression**.

8.1 LINEAR REGRESSION

Sir Francis Galton (1822–1911) thought that the mean height within a family of children should tend to equal the mean height of their parents. In other words that tall parents would produce tall children.

What he found was that the mean height of the children tended to 'regress' (go back) towards the mean population height. The name regression has stuck since to describe the relationship that exists between two or more **variables** (in Galton's case, parents' height and height of child; in the example above plant growth and amount of water).

To develop the idea further let us take the example of increasing activity affecting a person's percentage of body fat. First we have to have obtained appropriate permission for the **clinical trial** to be carried out. We have to assume that the people chosen are representative of the population we wish to study (say, women in a county who are both aged between 30 and 40 and are overweight). We will ask them to take different amounts of exercise and we will take blood samples monthly to record the concentration of a chemical which is known to reflect percentage body fat very well. We will select individual women at random, discuss the project with them and ask them if they are willing to take part. If so, we allocate ten at random to the lowest amount of exercise and ten each to receive each of the other three higher amounts.

Ideally, we do not inform the women whether they are taking a relatively high or low amount of exercise (this is known as a **blind trial**). This ensures that there is no psychological bias. Otherwise the knowledge that they were, for example, taking the greatest amount of exercise might subconsciously affect the manner in which they carried out the exercise and, hence, the results. In the same way, if you are recording the results of an experiment try to avoid thinking about what results you expect (want) from each treatment as you are recording it.

We can now work out how linear regression operates by developing a model as we did for analysis of variance (Chapter 6). First, imagine that we knew the concentration of the chemical in the blood at the end of the trial for each woman, but we did not know the amount of exercise which had been taken. The sum-of-squares (of the differences between the observations and their mean) can be calculated (Fig. 8.2, which shows only a few observations, for clarity). It represents the total amount of variability in our data.

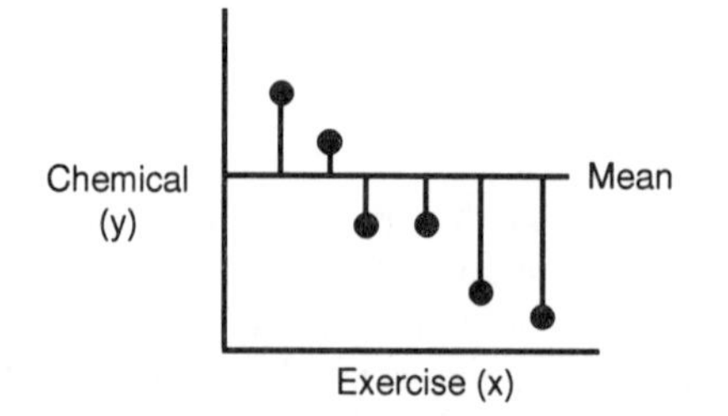

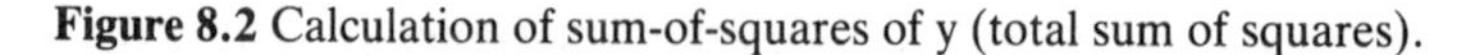
Figure 8.2 Calculation of sum-of-squares of y (total sum of squares).

8.2 THE MODEL

We now want to divide that variability into two parts: the amount that we can account for by knowing which amount of exercise the patient received and the rest (random variation). This is similar to dividing up the variation in analysis of variation into treatment and residual effects. However, we ask the question in a slightly different way. Specifically, if the plot of the data shows an approximately 'straight-line' relationship with a slope, how much of the total variability in 'concentration of chemical in the blood' (y) can be explained by a linear relationship with 'increasing amount of exercise' (x)?

If the real relationship is a close one (the amount of exercise does have a major effect on the concentration of the chemical in the blood) we will find that almost all of the variation in y can be accounted for by knowing x. In this case, the 'best-fit line' (the line that best describes the relationship) will be close to all the points on the graph (Fig. 8.3). If there are many other factors which may affect the concentration of the chemical in the blood (for example, age and body weight) then the scatter of points about the line will be much wider.

Each woman has taken a particular amount (x) of exercise and we have then noted the observed value of the chemical 'y'. The 'line of best fit' calculated to describe this relationship will enable us in future to read off an 'expected' or 'fitted' or 'predicted' value of y for any particular value of x. This is unlikely to be identical to the observed value of y for that value of x. The difference between the two is called a residual (Fig. 8.4), just as with analysis of variance. It tells us the discrepancy between our model (the straight line) and the data.

$$\text{Residual} = \text{observed y} - \text{fitted y}$$

A residual may be a positive or negative number. The method used to obtain the line of best fit is to minimize the sum-of-squares of the distances of the observations from the line, in a vertical direction. Another way of saying this is: to minimize the sum-of-squares of the residuals. This is

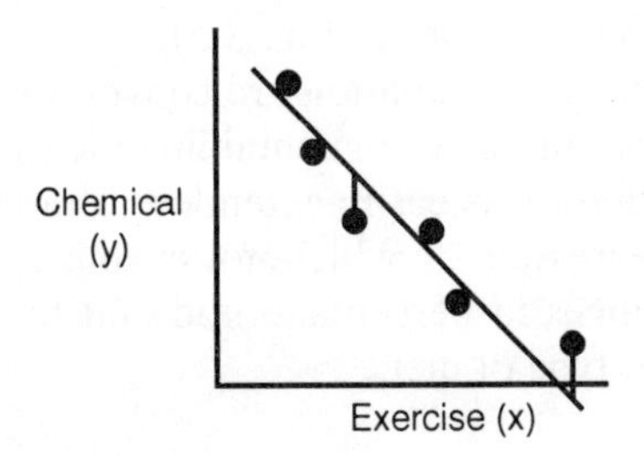

Figure 8.3 Best-fit line = regression line.

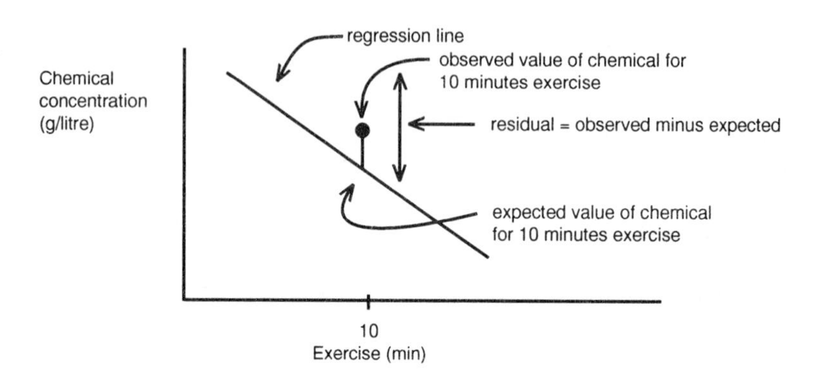

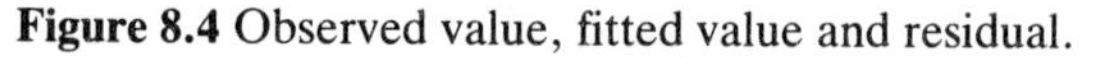

Figure 8.4 Observed value, fitted value and residual.

known as a 'least squares fit'. The vertical direction is used because we are trying to account for variation in y which is on the vertical axis.

DESCRIBING A STRAIGHT LINE BY AN EQUATION

First we put a y axis and an x axis on our graph. To specify a particular straight line we need to know two things: the value of y when x is zero (where the line cuts the y axis), and by how much the line goes up (or down) for each increase of one unit in x (the slope or gradient of the line).

You may already have met the equation of a straight line in mathematics as $y = mx + c$. In this form the letter c represents the value of y when x is zero and the letter m represents the slope. In statistics it is standard practice to use the letter a to represent the value of y when x is zero and the letter b to represent the slope. The equation is thus written:

$$y = a + bx$$

with a: estimate of the regression constant or intercept and b: estimate of the regression coefficient or slope (Fig. 8.5).

Notice that the terms of the equation are conventionally presented with the intercept preceding the term containing the slope. This latter symbolism is used because it is easily extended to more complicated models like multiple regression, in which, for example, we might wish to try and explain changes in percentage body fat by knowing both amount of exercise and type of diet.

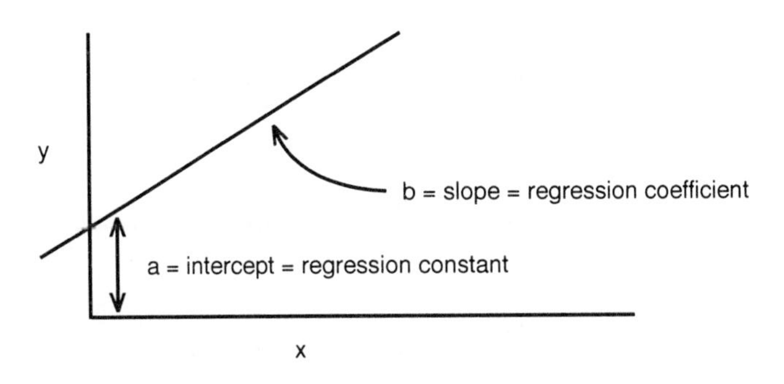

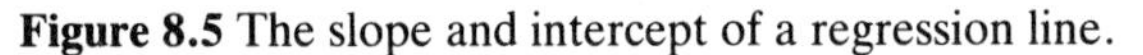
Figure 8.5 The slope and intercept of a regression line.

Let us use MINITAB to examine the data from our clinical trial. We have entered the x values (amount of brisk walking in minutes per day) and the y values (concentration of a chemical in the blood in grams per litre) into columns 1 and 2:

ROW	exercise	chemical
1	5	0.90
2	5	0.85
3	5	0.79
4	5	0.95
5	5	0.97
6	5	0.82
7	5	0.86
8	5	0.93
9	5	0.91
10	5	0.82
11	10	0.70
12	10	0.75
13	10	0.81
14	10	0.84
15	10	0.89
16	10	0.73
17	10	0.77
18	10	0.76
19	10	0.84
20	10	0.74
21	15	0.73
22	15	0.64
23	15	0.69
24	15	0.61
25	15	0.75
26	15	0.72
27	15	0.67
28	15	0.68

29	15	0.71
30	15	0.64
31	20	0.59
32	20	0.54
33	20	0.68
34	20	0.51
35	20	0.59
36	20	0.62
37	20	0.53
38	20	0.52
39	20	0.50
40	20	0.55

First we plot the observations to see whether they look as though a straight line with a slope is a sensible model:

```
MTB > plot c2 c1
            -
     +0.96  2
            -  *
 chemical - 2                 *
            -  2
            -  2                 2
     0.08+     *                 *
            -                    2
            -                    3             3
            -                    *             2
            -                                  2             *
     0.64+                                     2
            -                                  *             *
            -                                                2
            -                                                3
            -                                                3
     0.48+

       ---+------------+------------+------------+------------+------------+----exercise
         6.0          9.0          12.0         15.0         18.0        21.00
```

Since this looks reasonable (it resembles a straight line with a slope rather than either a random scatter of points or a curve) we ask for the regression line to be fitted and tested, with residuals being put into column 3 (so that we can ask for them to be plotted later). Note that a figure '1' appears between C2 and C1 to indicate that there is only one 'x' variable. The column containing the 'y' variable comes first:

MTB > regr c2 1 c1;
SUBC > resids c3.

First, this produces the equation of the line in the form y = a + bx:

The regression equation is
chemical = 0.990 – 0.0210 exercise

Notice that the coefficient (b) has a negative value here (–0.0210). This tells us that the line has a negative slope; it goes from top left to bottom right.

Second, MINITAB tests whether the intercept (a) and the slope (b) are significantly different from zero:

Predictor	Coef	Stdev	t-ratio	p
Constant	0.99000	0.02120	46.69	0.000
exercise	–0.021000	0.001549	–13.56	0.000

This printout shows that the intercept is significantly different from zero (Constant, Coef, p = 0.000 . . ., which is much less than 0.05). Therefore, when a woman takes no exercise this regression line predicts that the concentration of the chemical in her blood will be significantly greater than zero – the estimated value is 0.99 g per litre. Also the slope of the line (exercise, Coef) is significantly different from zero (p = 0.000 again). We can therefore be very confident that there is a strong linear relationship between chemical concentration and exercise.

Third, MINITAB gives the standard deviation (s = square root of the residual or error variance) and 'r squared' (R-sq):

s = 0.05475   R-sq = 82.9%   R-sq(adj) = 82.4%

R squared is also commonly referred to as r^2 and is called the **coefficient of determination**. It is the proportion of the variation in y accounted for by variation in x (from 0 to 1, which is the same as from 0 to 100%). If r squared is 100% it indicates that the regression line goes through all the points on the graph. Our model then explains all the variability in the response variable: there is no random variation, all the residuals are zero. In contrast, if r squared is 0% it is consistent with a random arrangement of points on the graph. The larger the value of r squared, the more useful the independent variable is likely to be as a predictor of the response variable. (Don't worry about the adjusted r squared value on the printout. It is only helpful in more complicated models.)

Fourth, the analysis of variance table appears. This is of the same general form as for when we are comparing the effects of several treatments in an experiment. However, here the treatments line is replaced by one called 'regression'. This represents the evidence for a linear relation between the chemical and exercise.

Analysis of variance

SOURCE	DF	SS	MS	F	p
Regression	1	0.55125	0.55125	183.91	0.000
Error	38	0.11390	0.00300		
Total	39	0.66515			

The evidence is very strong (p = 0.000, which is much less than 0.05). We can reject the null hypothesis of no linear relationship between chemical concentration and amount of exercise with great confidence. MINITAB is carrying out the same test here as the test shown above for whether the slope of the line is significantly different from zero (the slope of a horizontal line). We can see this by noting that the value of the variance ratio (or F, 183.91) is the square of the ratio given above for the slope (–13.56). This will always be the case. It is probably easier to concentrate on the analysis of variance when interpreting the analysis.

Notice that the value of r squared (82.9% or 0.829) can be obtained by dividing the regression sum-of-squares by the total sum-of-squares.

Next MINITAB notes any observations which are 'unusual':

Unusual Observations

Obs.	exercise	chemical	Fit	Stdev.Fit	Residual	St.Resid
15	10.0	0.89000	0.78000	0.00948	0.11000	2.04R
33	20.0	0.68000	0.57000	0.01449	0.11000	2.08R

R denotes an obs. with a large st. resid.

This tells us that observations numbers 15 and 33 have rather large residuals. This can be useful in helping us to spot copying errors in our data. We should always check such values. Here we will assume that we find that they are correct.

In interpreting the output it is important to realize that, just because you have shown a significant relationship, this does not automatically mean that x is **causing** the variation in y. Both could be responding independently to some third factor (z) (see later). Also the model makes assumptions about the nature of the data. If these assumptions are not correct then the interpretation may again be invalid.

8.3 ASSUMPTIONS

1. A **normal distribution** of residuals is assumed, since they represent random variation. There should be many residuals with very small absolute values (near zero) and only a few with very large ones (far from zero). We can test this by asking for a histogram of the residuals in MINITAB which we asked to be stored in column 3:

```
MTB > hist c3
Histogram of C3 N = 40
Midpoint   Count
 -0.10       1     *
 -0.08       1     *
 -0.06       5     *****
 -0.04       7     *******
 -0.02       5     *****
  0.00       3     ***
  0.02       5     *****
  0.04       4     ****
  0.06       5     *****
  0.08       2     **
  0.10       0
  0.12       2     **
```

We see that the residuals have an approximately normal distribution (albeit a bit 'flat').

2. **Homogeneity of variance** of residuals is assumed. The residuals should not show a tendency to increase (or decrease) as x increases. We can examine this by plotting residuals against values of x in MINITAB.

```
MTB > plot c3 c1

residuals
         -                      *                                   *
         -
         -    *
  0.070+      *                                *
         -                      2              *                    *
         -    *                                *
         -    *                 *              *
         -    *                                *                    2
  0.000+                                       2
         -                      2                                   *
         -   2                  *                                   *
         -                      *              2                    *
         -                      *                                   2
 -0.070+     2                                 *                    *
         -                      *
         -    *
         -
        ---+------------+------------+------------+------------+------------+----exercise
          6.0          9.0         12.0         15.0         18.0        21.00
```

The variability in the residuals is similar for all four amounts of exercise. This is fine. There is no obvious pattern. A common problem is shown in Fig. 8.6 where there is much more variability at high levels of x than at low levels. If this occurs consult a statistician.

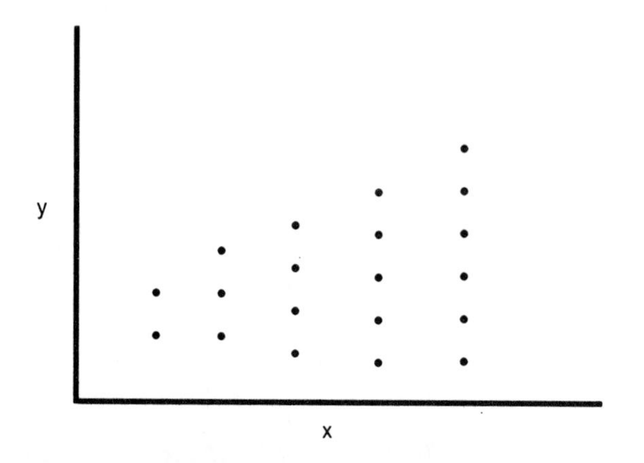

Figure 8.6 Variability in y increases with x. This is the 'shotgun effect'.

3. **Independence** Every observation of 'y' (chemical concentration in the MINITAB example) must be independent of every other 'y'. If, for example, we included several women from the same household in our clinical trial their observations will not be independent of each other because the women may well have the same diet and be closely genetically related to each other.

We need to ensure independence by carrying out our sampling or experimentation in an appropriate way. For example, in an experiment to study the effect of water on plant growth each observation should come from a plant growing in a separate pot.

4. **Continuity** Ideally both x and y variables should be measured on a continuous scale (like kilograms, minutes or centimetres) rather than being counts, proportions or ranks. Counts and proportions can also often be analysed using regression. It is important to examine the graph of the residuals and, if in any doubt, you should ask advice.

Other special techniques for ranked observations are introduced in Chapter 9.

5. **Absence of 'errors' in values** Ideally the x observations should be 'without error'. Obviously we want our measurement of the x values to be as accurate as possible although we realize that they will not be perfectly correct. If there is no bias slight mistakes either way in measuring x values will merely increase the residual mean square and make the regression less precise. However, if there is reason to suspect bias in the observation of x such that it is generally recorded as too high (or generally recorded as too low) this will affect the estimation of the slope.

8.4 FURTHER EXAMPLE OF LINEAR REGRESSION

In an experiment such as the clinical trial described above we usually have only three of four levels of the independent variable (x) and several replicate results for each level. However, the principles of regression analysis extend to situations where x values are not evenly spread and where there may be only one sample per x value. The regression process is still the same – we need to calculate the line which minimizes the sum-of-squares of the points from the line in a vertical direction.

MINITAB will carry out the required calculations on the pairs of values for each individual and produce an equation which represents the regression line.

Let us take an example of data from a survey in which the amount of lead in parts per million (y) on the vegetation alongside 19 roads depends on the amount of traffic (hundreds of vehicles passing per day) (x).

We can enter the data into columns 1 and 2 of a MINITAB worksheet:

ROW	lead	traffic
1	44	79
2	58	109
3	43	90
4	60	104
5	23	57
6	53	111
7	48	124
8	74	127
9	14	54
10	38	102
11	50	121
12	55	118
13	14	35
14	67	131
15	66	135
16	18	70
17	32	90
18	20	50
19	30	70

Now examine the following printout and interpret it:

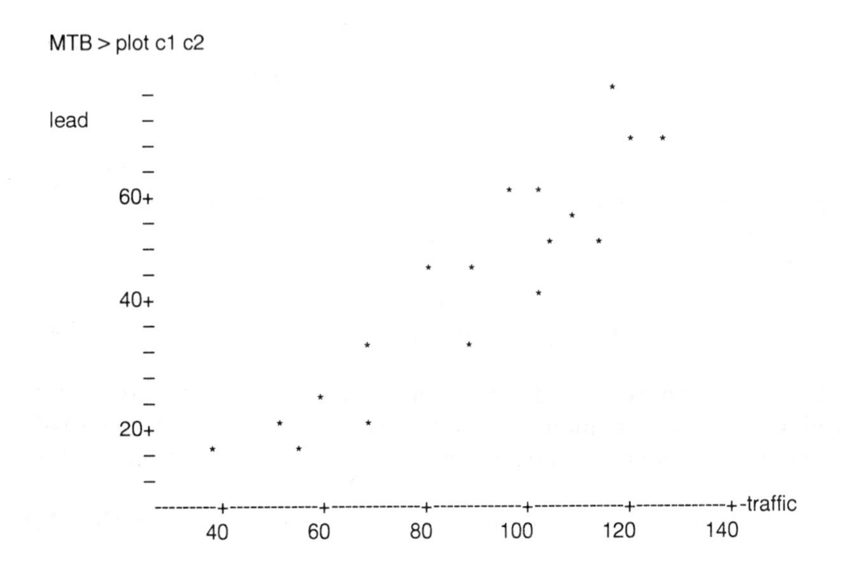

MTB > regress cl 1 c2;
SUBC > resids c3.

The regression equation is
lead = – 10.7 + 0.569 traffic

What is the intercept? What is the slope?

Predictor	Coef	Stdev	t-ratio	p
Constant	–10.736	5.813	–1.85	0.082
traffic	0.56893	0.05923	9.61	0.000

Is the intercept significantly different from zero? (Note that the estimate of –10.736 must be meaningless since we cannot have a negative value for lead. It may well be that if our sample included lower levels of traffic we would find that the relationship was curved (Fig. 8.7).)

Is the slope significantly different from zero?

Is this equation of a straight line a useful model to describe the relationship between roadside lead and traffic numbers?

s = 7.680   R-sq = 84.4%   R-sq(adj) = 83.5%

What proportion of the variation in lead concentration is accounted for by amount of traffic?

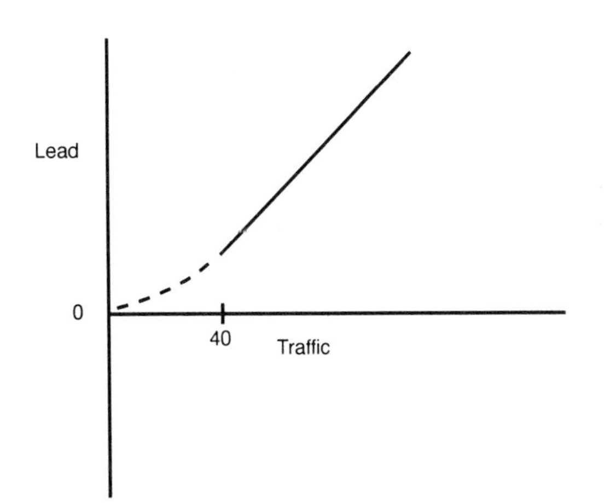

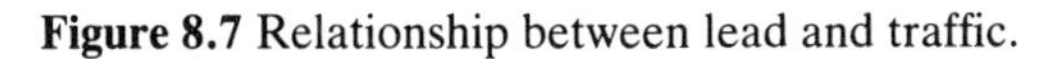

Figure 8.7 Relationship between lead and traffic.

Analysis of Variance

SOURCE	DF	SS	MS	F	p
Regression	1	5442.0	5442.0	92.26	0.000
Error	17	1002.8	59.0		
Total	18	6444.7			

With what confidence can you reject the null hypothesis of no linear relationship between lead concentration and amount of traffic?

```
MTB > hist c3
Histogram of resids N = 19
Midpoint   Count
   -12       2     **
    -8       3     ***
    -4       1     *
     0       5     *****
     4       4     ****
     8       2     **
    12       2     **
```

Do the residuals show an approximately normal distribution?

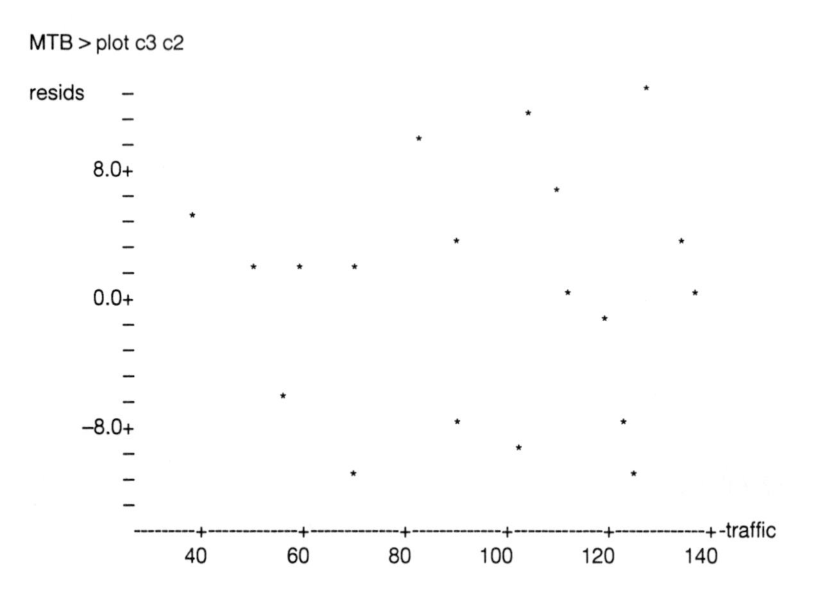

Do the residuals show similar variability over the range of x values?

If you have release 8 of MINITAB and high resolution graphics you can now obtain a neat plot of the observations with the regression line superimposed (Fig. 8.8).

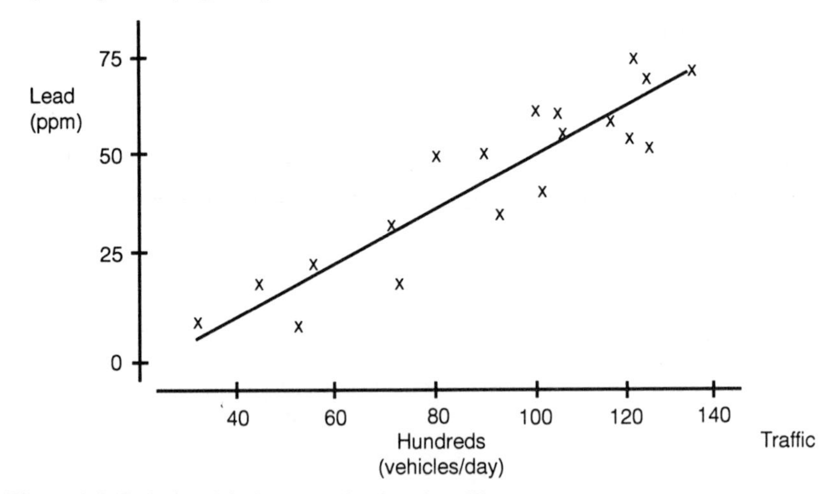

Figure 8.8 Relationship between lead and traffic.

8.5 THE IMPORTANCE OF PLOTTING OBSERVATIONS

The lead deposition/traffic dataset which we have just analysed was a typical example of a relationship which is well modelled by linear regression. However, it is possible for the computer to give you the same equation (and the same values for p and r squared) from very different datasets for which the model would be inappropriate. This is an alarming thought. It is important to plot the observations to identify such problems.

We will use four datasets devised by Anscombe* to illustrate this problem. The x values for datasets 1, 2 and 3 are the same and are in column 1 while that for dataset 4 is in column 5. The corresponding responses are in columns named y1, y2, y3 and y4:

ROW	x123	y1	y2	y3	x4	y4
1	10	8.04	9.14	7.46	8	6.58
2	8	6.95	8.14	6.77	8	5.76
3	13	7.58	8.74	12.70	8	7.71
4	9	8.81	8.77	7.11	8	8.84
5	11	8.33	9.26	7.81	8	8.47
6	14	9.96	8.10	8.84	8	7.04
7	6	7.24	6.13	6.08	8	5.25
8	4	4.26	3.10	5.39	19	12.50
9	12	10.84	9.13	8.15	8	5.56
10	7	4.82	7.26	6.42	8	7.91
11	5	5.68	4.74	5.73	8	6.89

8.5.1 A straight line relationship with some scatter

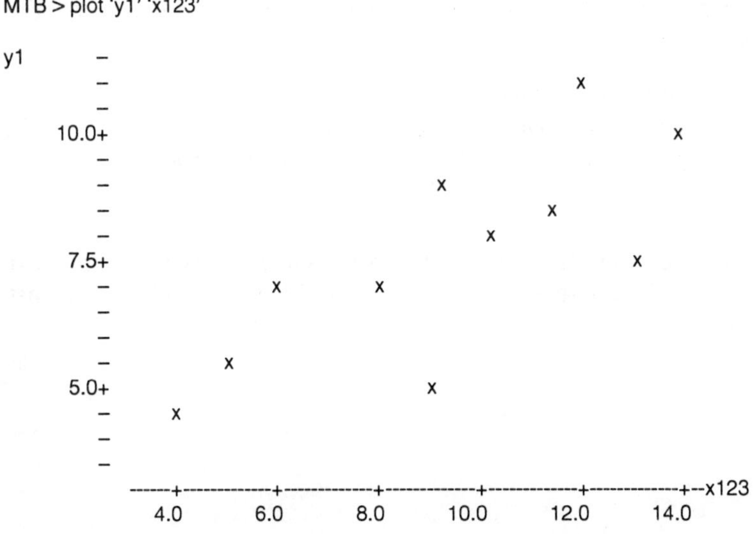

* Anscombe, F.J. (1973) Graphs in statistical analysis, *American Statistician*, **27**, 17–21.

In carrying out the regression we can ask for MINITAB to calculate fitted values and automatically put them into a column called 'FITS'. It will call the fitted values from the first analysis 'FITS1' and those from the next analysis 'FITS2'. Instead of asking for ordinary residuals we have asked for what are called 'standardized residuals' to be calculated and put automatically into a column named 'SRES1'. The values of ordinary residuals will depend on the units in which y was measured. If we ask for them to be standardized by dividing them by an estimate of their variability we can judge their size more easily. If we find that a standardized residual is greater than 2 or less than –2 we should be alert to possible errors in the data or to the possibility that we are not fitting a sensible model.

MTB > Regress 'y1' 1 'x123' 'SRES1' 'FITS1'.

The regression equation is
y1 = 3.00 + 0.500 x123

Note that the values of the intercept (3.0) and the slope (0.5) is the same in this and the next three analyses. The regression line fitted is exactly the same for **all four** datasets.

Predictor	Coef	Stdev	t-ratio	P
Constant	3.000	1.125	2.67	0.026
x123	0.5001	0.1179	4.24	0.002
s = 1.237	R-sq = 66.7%	R-sq (adj) = 62.9%		

Note that r-squared (66.7%) is the same in this and in the next three analyses.

Analysis of Variance

SOURCE	DF	SS	MS	F	P
Regression	1	27.510	27.510	17.99	0.002
Error	9	13.763	1.529		
Total	10	41.273			

Note that the probability (p = 0.002) is the same in this and in the next three analyses. We now ask for standardized residuals to be plotted against fitted values:

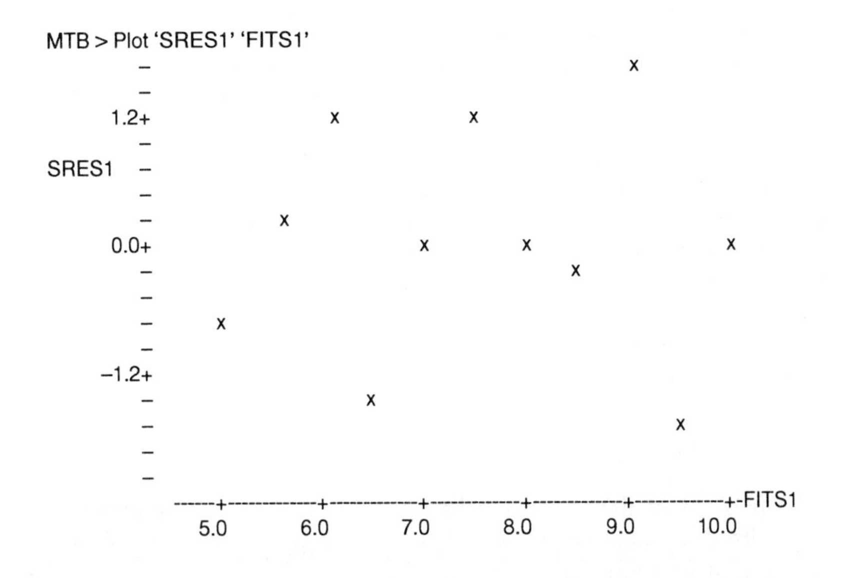

There is no pattern in the residuals. A straight-line model seems satisfactory.

8.5.2 A curved relationship

Although the response increases from left to right, the rate is not steady, the response levels off. Here, the fit of the regression line is poor. Predicted values are too high at extreme values of x and too low in the middle.

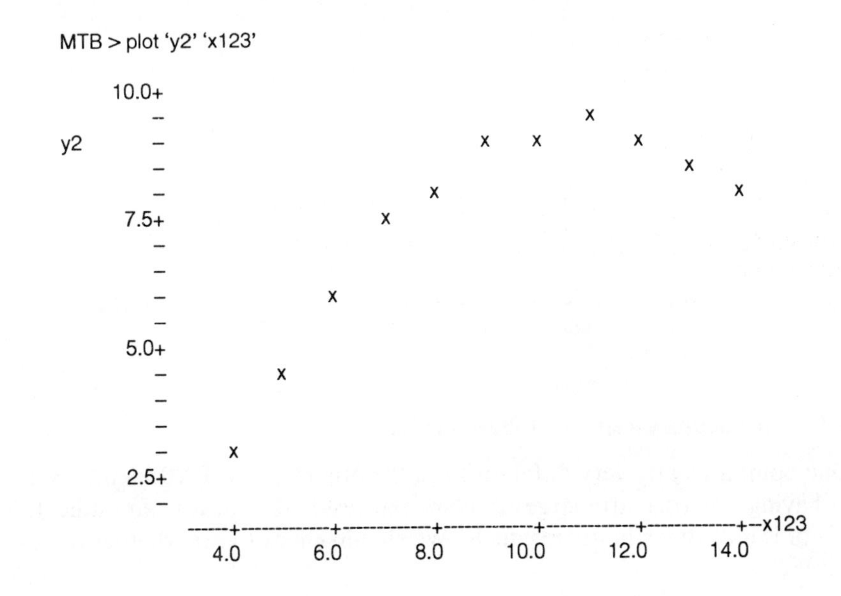

MTB > Name c9 = 'SRES2' c10 = 'FITS2'
MTB > Regress 'y2' 1 'x123' 'SRES2' 'FITS2'.

The regression equation is
y2 = 3.00 + 0.500 x123

Predictor	Coef	Stdev	t-ratio	p
Constant	3.001	1.125	2.67	0.026
x123	0.5000	0.1180	4.24	0.002

s = 1.237 R-sq = 66.6% R-sq(adj) = 62.9%

Analysis of Variance

SOURCE	DF	SS	MS	F	p
Regression	1	27.500	27.500	17.97	0.002
Error	9	13.776	1.531		
Total	10	41.276			

The poor fit of the regression line is confirmed by the pattern of the standardized residuals when plotted against the fitted values. They are not randomly distributed. They are positive in the middle and negative at either end of the range of fitted values.

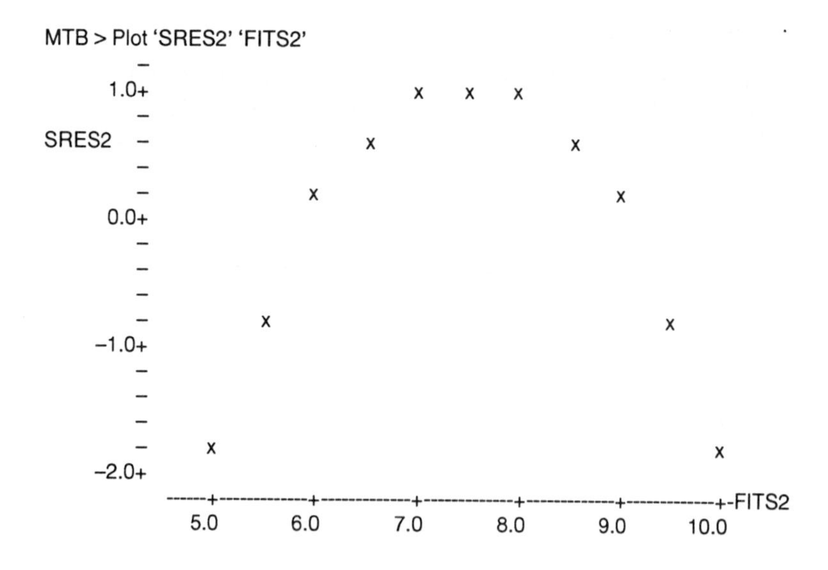

8.5.3 All points except one follow one line

One point is clearly very different from the others. MINITAB identifies it as having a particularly large standardized residual (3.0, it is an outlier). There is no pattern in the residuals. We should check to see whether there is an error in the data. If the data point is correct subsequent analysis and interpretation both with and without the point seems sensible.

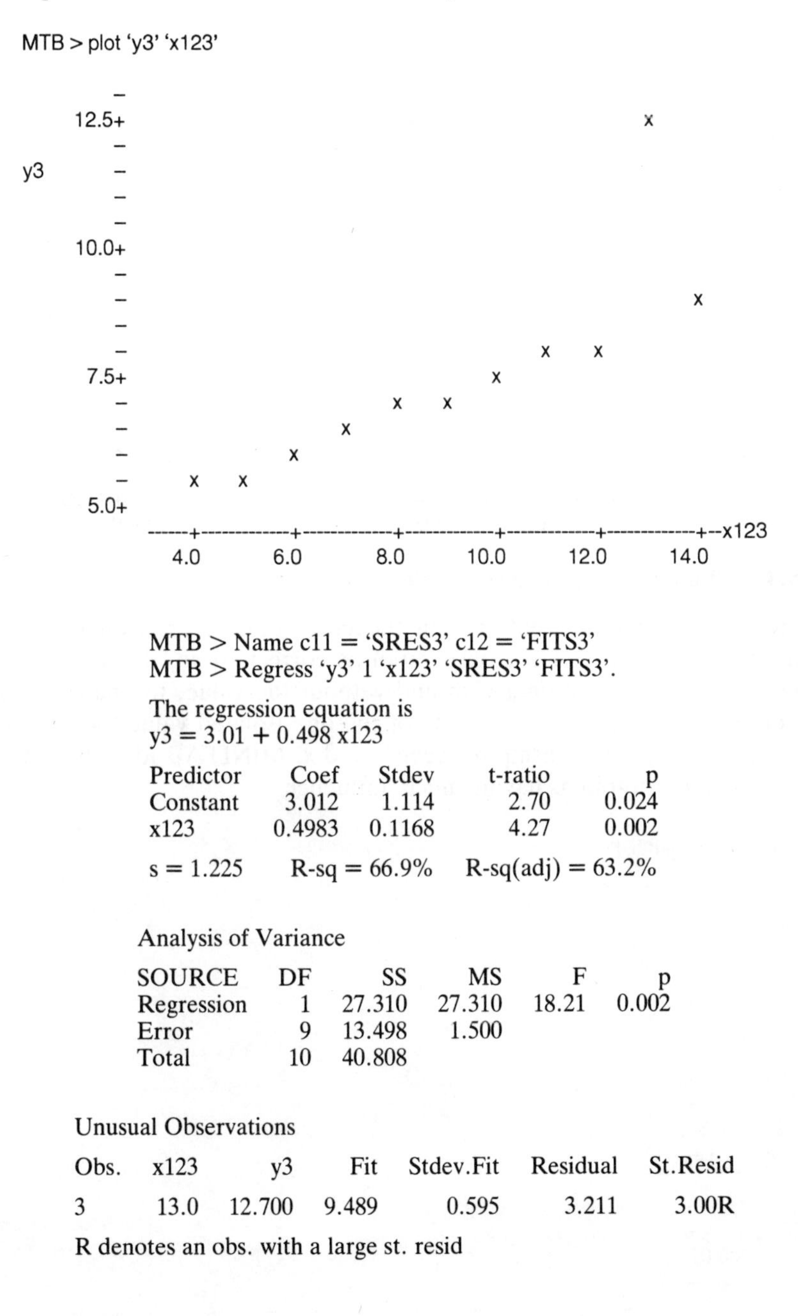

MTB > Name c11 = 'SRES3' c12 = 'FITS3'
MTB > Regress 'y3' 1 'x123' 'SRES3' 'FITS3'.

The regression equation is
y3 = 3.01 + 0.498 x123

Predictor	Coef	Stdev	t-ratio	p
Constant	3.012	1.114	2.70	0.024
x123	0.4983	0.1168	4.27	0.002

s = 1.225 R-sq = 66.9% R-sq(adj) = 63.2%

Analysis of Variance

SOURCE	DF	SS	MS	F	p
Regression	1	27.310	27.310	18.21	0.002
Error	9	13.498	1.500		
Total	10	40.808			

Unusual Observations

Obs.	x123	y3	Fit	Stdev.Fit	Residual	St.Resid
3	13.0	12.700	9.489	0.595	3.211	3.00R

R denotes an obs. with a large st. resid

The plot of standardized residuals against fitted values clearly identifies the same problem.

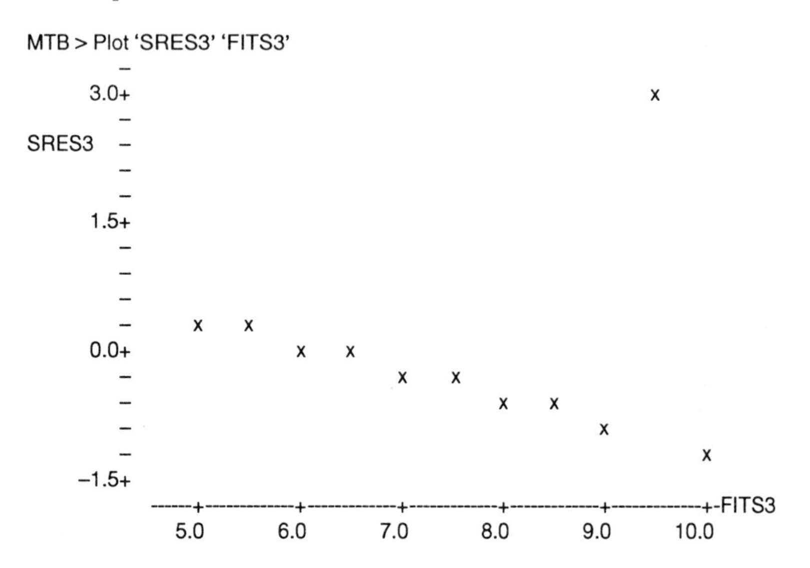

8.5.4 All points except one are clustered together

Here, one observation is determining the nature of the regression line. It is important to check that the observation is correct. If so, it again seems sensible to analyse the data with and without the value, to quantify the difference it makes. In the absence of the one isolated value there is no significant linear relationship between y and x. MINITAB identifies such points on the printout as having much 'influence'.

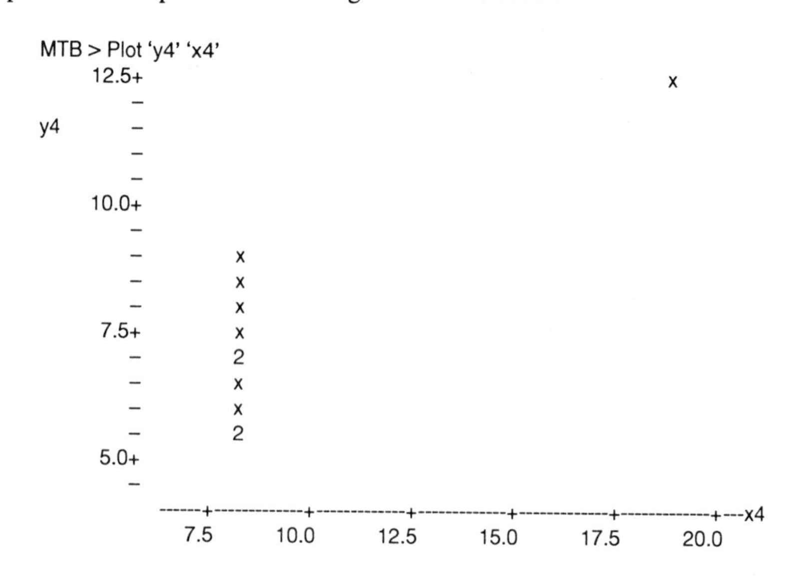

MTB > Name c13 = 'SRES4' c14 = 'FITS4'
MTB > Regress 'y4' 1 'x4' 'SRES4' 'FITS4'.

The regression equation is
y4 = 3.00 + 0.500 x4

Predictor	Coef	Stdev	t-ratio	p
Constant	3.002	1.124	2.67	0.026
x4	0.4999	0.1178	4.24	0.002

s = 1.236 R-sq = 66.7% R-sq(adj) = 63.0%

Analysis of Variance

SOURCE	DF	SS	MS	F	p
Regression	1	27.490	27.490	18.00	0.002
Error	9	13.742	1.527		
Total	10	41.232			

Unusual Observations

Obs.	x4	y4	Fit	Stdev.Fit	Residual	St.Resid	
8	19.0	12.500	12.500	1.236	0.000	*X	

X denotes an obs. whose X value gives it large influence.

The plot of standardized residuals against fitted values omits the value derived from the eighth because it is so far away from the others.

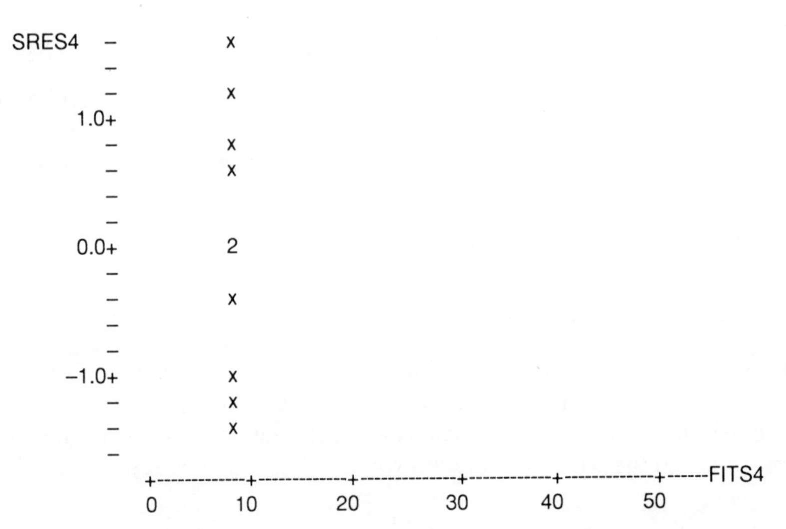

8.5.5 The difference between using linear regression in an experiment and in a survey.

The values of x are **imposed** in an experiment: for example we control the amount of exercise which 40 women take. A random group of ten walk briskly for 5 minutes per day, ten more for 10 minutes, ten more for 15 minutes and the final group of ten for 20 minutes. In contrast, in a survey we are not in control of the x values. They are whatever they happen to be. In both cases our conclusions must be restricted to the range of x values we have included in our study. But, whereas in the experiment we have evidence to show that a change in the amount of x causes a change in the amount of y, in a survey it may simply be that x and y are linearly associated because they are both related to a third variable.

8.5.6 Extrapolation is dangerous

If we predict a y value for an x value outside our range of data (**extrapolation**) we cannot be sure that the relationship will still be of the same form. For example a plant which is growing steadily faster over the range of fertilizer from 0 to 300 kg/ha is unlikely to continue growing at the same rate if fertilizer were applied at the rate of 1000 kg/ha – it will probably show signs of toxicity!

Therefore we should not extrapolate from our data; rather we should extend the range of treatments in a subsequent experiment.

8.6 CONFIDENCE INTERVALS

A regression line is an estimate of the relationship between y and x. If earlier we had taken a different random sample of roads or plants our estimated line would have been somewhat different. For example we might have obtained

$$\text{lead} = -9.90 + 0.623\ \text{traffic}$$

8.6.1 For the slope of the line

As with the estimate of a mean (Chapter 2) we should consider calculating the 95% confidence interval for the slope of the line. This gives us the range within which we are 95% confident that the slope of the line appropriate to the whole population will lie. These can be represented by:

$$b \pm t \times \sqrt{\frac{\text{RMS}}{\text{Sxx}}}$$

with t taken from tables for p = 0.05 and for n-2 df (where n is the number of points on the graph); where RMS = residual (or error) mean square and Sxx = the sum-of-squares of the x values.

This equation is of the same general form as that for a population mean (Chapter 2) in that we have an estimated value plus or minus t times the standard error of that value.

8.6.2 For the regression line

We often subsequently wish to predict the value of y (y′) for a particular individual with a value of x (x′). This can be achieved as follows:

$$y' \pm t \times \sqrt{\text{RMS} \times \left(1 + \frac{1}{n} + \frac{(x' - \bar{x})^2}{\text{Sxx}}\right)}$$

with t taken from the tables for n-2 df and an appropriate probability level (where n = number of points on the graph); where RMS = residual (or error) mean square; Sxx is the sum-of-squares of the x values and $\bar{x}$ is the mean of the x values.

Again this follows the format: estimate plus or minus t times the standard error of that estimate. It is just that here the standard error is more complex.

If we work out confidence intervals for three values of x (Fig. 8.9a) we can join them together to provide a confidence interval for the line (Fig. 8.9b). We are 95% confident that the line which represents the relationship for the entire population is within this range. Notice how the confidence interval is narrowest in the centre of the graph and becomes wider at the extremes. This is because outside the region of our observations we know nothing about the relationship and so as we approach this area we become less and less certain about the location of the line.

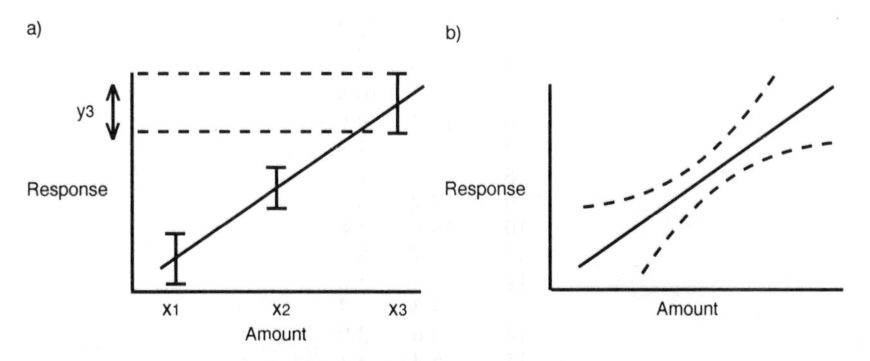

Figure 8.9 Confidence interval for (a) a particular value of y; (b) the regression line.

8.7 CORRELATION

So far in this chapter we have worked with data where it is clear that one variable may depend on the other. Sometimes however we may simply wish to know to what extent two variables of equal status increase or decrease with each other in a straight line relationship. We are not concerned with predicting one from the other but we want to know if they are varying in a related way, i.e. are **correlated**.

For example, the concentration (ppm) of two chemicals in the blood might be measured from a random sample of 12 patients suffering to various extents from a particular disease. If a consequence of the disease is that both chemicals are affected we should expect patients with high values of one to have high values of the other and *vice versa*. The extent of the linear correlation between the chemicals can be estimated by calculating a correlation coefficient (r). R can be positive or negative, ranging between –1 and +1. A value near +1 indicates a strong positive linear correlation while a value near –1 shows a strong negative relationship. A value of 0 may either indicate randomness or that a more complicated relationship is present (a curve, for example); a plot helps here. R is the square root of r^2, the coefficient of determination which we saw in the printout from regression analysis. Let us examine a set of data.

The concentrations of chemical A in the blood are in C1 of a MINITAB worksheet and the concentrations of chemical B are in C2. Both are in micrograms per litre. There is one row for each of 15 patients.

```
MTB > print c1–c2

ROW     A      B
  1   46.1   24.7
  2   23.6   15.2
  3   23.7   12.3
  4    7.0   10.9
  5   12.3   10.8
  6   14.2    9.9
  7    7.4    8.3
  8    3.0    7.2
  9    7.2    6.6
 10   10.6    5.8
 11    3.7    5.7
 12    3.4    5.6
 13    4.3    4.2
 14    3.6    3.9
 15    5.4    3.1
```

First, we plot the data:

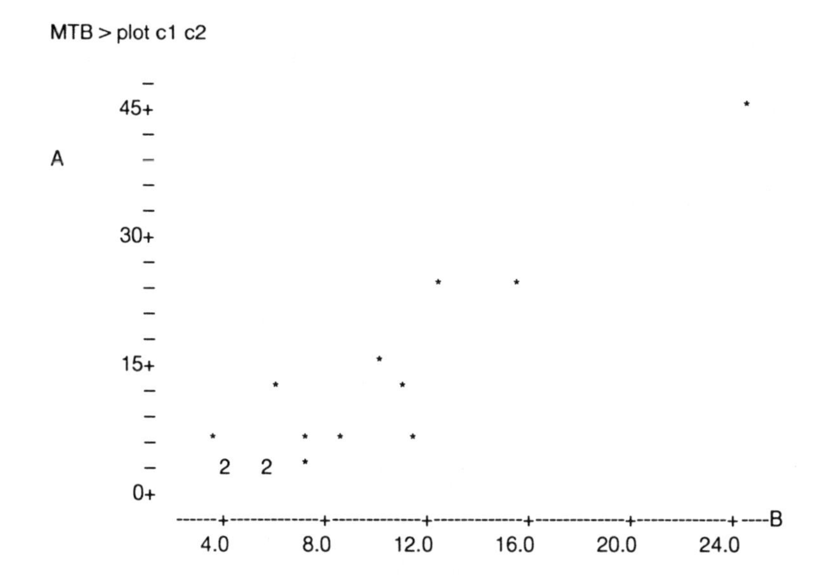

We can ask for the correlation coefficient, r:

MTB > corr c1 c2
Correlation of A and B = 0.939

This is very high, suggesting a strong, positive linear correlation. We can find out whether this value of r indicates a significant correlation by comparing it with a critical value found in a statistical table of the linear correlation coefficient, r. The null hypothesis is that the correlation coefficient is zero. If our estimate of it (which we make positive if it happens to be negative, as it is only the size of the coefficient rather than its sign which is important) is greater than the value found in the table for the appropriate n (number of points on the graph) and $p = 0.05$ we conclude that there is a significant correlation. In this case the critical value for $p = 0.05$ is 0.514, while for $p = 0.01$ it is 0.641, so we have strong evidence for a significant correlation.

However, the point at the top right of the plot may well have a considerable influence on this result. We could simply run the analysis with and without the point and compare the results but by omitting the result we are not making as much use of the data as we might. After all it is (we believe) a real result. An alternative is therefore to opt to use a different correlation coefficient: Spearman's r. (The 'classical' one which we have been using up to now can similarly be called Pearson's r, or, sometimes, the product-moment correlation coefficient. This is the square root of the r-squared value we met earlier in linear regression.)

Spearman's r is calculated using the ranks of the observations as follows. The lowest value for A is 3.0 so this has a rank of 1, while the highest value is 46.1, so this has a value of 15. The amounts of B are also ranked, with 3.1 receiving a rank of 1 and 24.7 a rank of 15. MINITAB can do this for us and put the results in columns 3 and 4:

MTB > rank c1 c3
MTB > rank c2 c4

ROW	A	B	rA	rB
1	46.1	24.7	15	15
2	23.6	15.2	13	14
3	23.7	12.3	14	13
4	7.0	10.9	7	12
5	12.3	10.8	11	11
6	14.2	9.9	12	10
7	7.4	8.3	9	9
8	3.0	7.2	1	8
9	7.2	6.6	8	7
10	10.6	5.8	10	6
11	3.7	5.7	4	5
12	3.4	5.6	2	4
13	4.3	4.2	5	3
14	3.6	3.9	3	2
15	5.4	3.1	6	1

We can now ask for a correlation coefficient to be calculated for the ranked data (Spearman's r):

Spearman's Rank Correlation coefficient
MTB > corr c3 c4
Correlation of rA and rB = 0.764

The outlier clearly had a considerable effect on the correlation coefficient because the 'r' value has come down to 0.764. We can see the difference by plotting the ranked data:

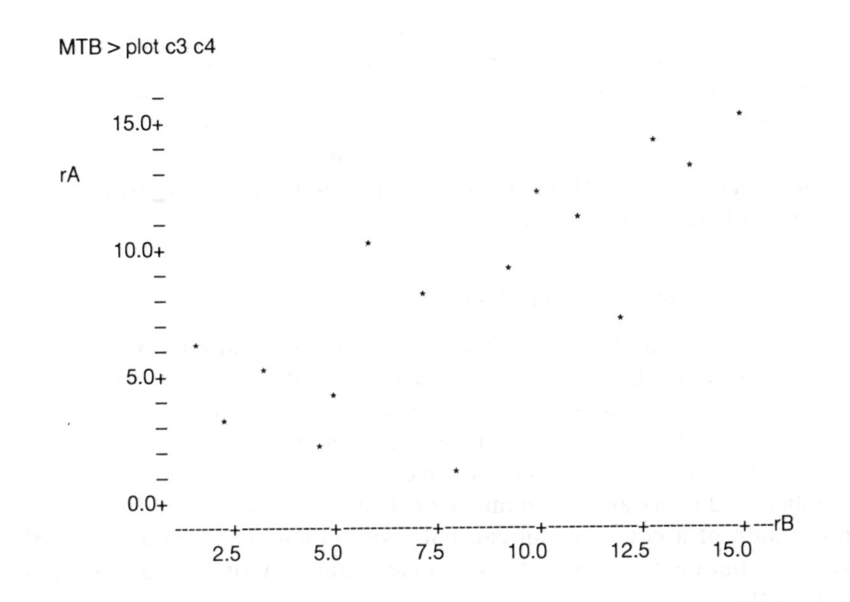

This examines whether there is a tendency for the two components of the matched pairs to increase and decrease together (or for one to increase as the other decreases). Ranking the data maintains the order of the observations on each axis. However, the size of the difference between one observation and the next biggest one is standardized. Now the patient who was an outlier is still the point at the top right of the graph but, because the data are now ranked, he or she is not so far removed from the other data points as before. Spearman's r is described as being more robust because it is less sensitive than Pearson's r to occasional outliers or to bends in the relationship.

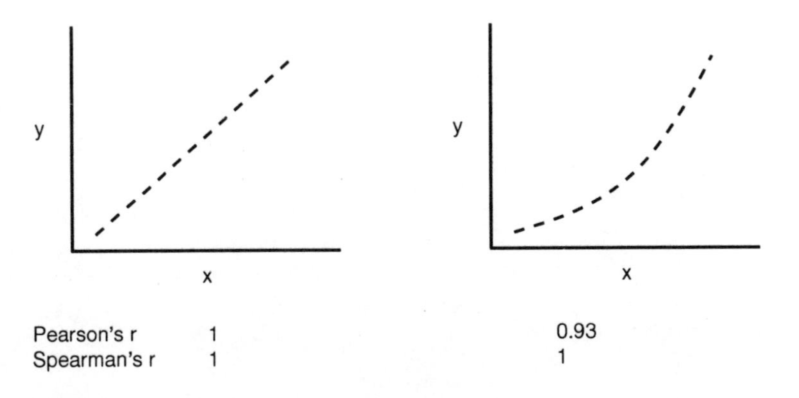

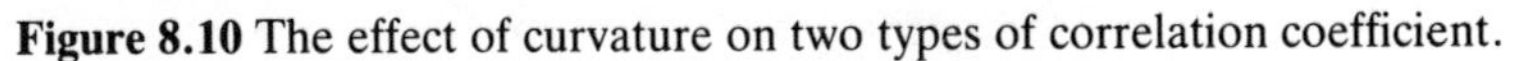
Figure 8.10 The effect of curvature on two types of correlation coefficient.

Figure 8.10 illustrates this for two datasets showing how the value of Spearman's r remains at 1 for a positively (or negatively) curved relationship whereas Pearson's r gives a lower value, because of the lack of linearity.

Again we can test the significance of Spearman's r. A table of critical values gives one of 0.521 for $p = 0.05$ and 0.654 for $p = 0.01$. So we have strong evidence of correlation.

8.7.1 Correlation is not causation

It is commonly found that there is a strong positive linear relation between the number of doctors in a city and the number of deaths per year in that city! At first sight we may be tempted to conclude that having more doctors leads to more deaths. Therefore if we cut the number of doctors we might expect fewer deaths. However, we have overlooked the fact that both the number of doctors and the number of deaths in a city depend upon the population of a city. We can calculate correlations between any pair of variables but we must always be wary of assuming that one causes variation in the other.

What to do when data are skewed or are ranks or scores or are counts in categories

9

9.1 INTRODUCTION

Most of the statistical techniques we have considered so far are **parametric** tests (t-tests, analysis of variance and linear regression). This means that they make certain assumptions about parameters of the distribution of the population(s), e.g. that the data come from populations which are normally distributed and have similar variability. If data do not conform to these assumptions we need to use **non-parametric** or **distribution-free** tests instead. These do not make such strict assumptions about the shape of the population distributions.

The method of box-and-whisker plots (Chapter 2) which calculates a median and summarizes data in a way which highlights any departures from the assumptions for parametric tests provides a useful starting point for considering which techniques may be helpful.

Non-parametric tests are especially useful when:

1. Underlying distributions are not normal (for example they may be skewed – showing 'straggle' to the right or to the left) (Fig. 9.1a).

   It may be possible to modify the data so that parametric tests can be used. For example, if leaf areas have been measured we could analyse the square root of each value, since this provides an alternative measure of leaf size. This is called **transformation**. However, if we wish to analyse the raw data we need to consider non-parametric tests.
2. There are extreme observations (outliers) (Fig. 9.1b).
3. Observations are merely ranked or arbitrarily scored. For example, first, second, third or a score of 0 to 10 for depth of colour.

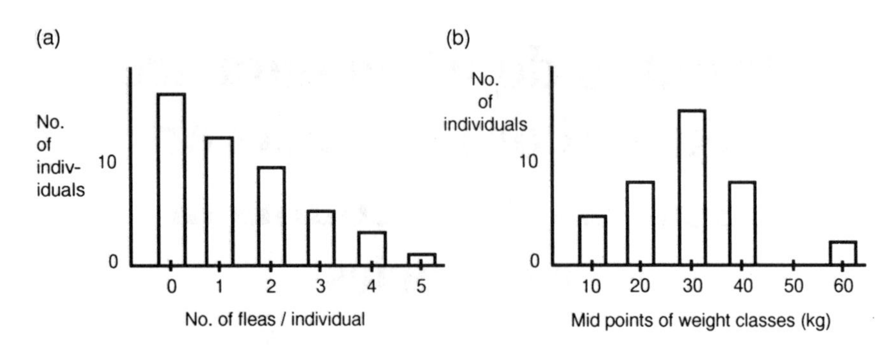

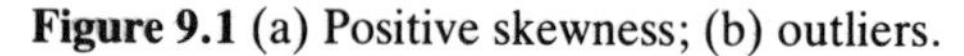

Figure 9.1 (a) Positive skewness; (b) outliers.

In the last chapter we described Spearman's Rank Correlation Coefficient, r_S which is a non-parametric method. We can use this to estimate the correlation between two variables when our data fail to meet the assumptions for using the equivalent parametric method (Pearson's Product Moment Correlation Coefficient: r).

In this chapter we will consider techniques which are alternatives to the parametric tests which we have already used. These are:

1. **Mann-Whitney** (sometimes also known as the Wilcoxon rank sum test). This is the non-parametric equivalent of analysis of variance when there are two treatments.
2. **Kruskal-Wallis**. This is the equivalent of one-way analysis of variance when there are two or more treatments. This can also cope with data of type (1).
3. **Friedman**. This is the equivalent of two way analysis of variance when there are blocks as well as two or more treatments.

In addition, we will discuss the **Chi-squared contingency** test. This is used for **categorical** data. Categorical refers to the fact that the data are the number of individuals in each treatment which have been classified into each of two or more categories (like dead or alive, rich or poor) rather than being measures of some feature. For example, we have transplanted 200 plants of one species into a field: 100 with red flowers and 100 with white flowers. One year later we can categorize both the red- and white-flowered plants as either living or dead. We might wish to know whether the red-flowered plants were more or less likely to survive than the white-flowered ones.

9.2 MANN-WHITNEY TEST

If we have an experiment with two treatments in which our data are either not measured on an absolute scale (for example they are scores or ranks rather than centimetres or kilograms) or they are skewed rather than normally distributed we should use a Mann-Whitney test. This tests whether there is a difference between the population medians. The assumptions we must be able to make are, first, the observations must be random and independent observations from the populations of interest; and second, the samples we compare are assumed to come from populations which have a similar shaped distribution (for example, both are positively skewed).

The growth form of young trees grown in pots can be scored on a scale from 1 (poor) to 10 (perfect). This score is a complex matter as it summarizes the height, girth and straightness of the tree. We have 16 trees. Eight of the trees were chosen at random to be grown in a new compost while the remaining eight were grown in a traditional compost. Unfortunately, one of the latter was accidentally damaged, leaving only seven trees. The null hypothesis is that the two composts produce trees with the same median growth form. The alternative hypothesis is that the composts differ in their effect.

After 6 months growth the scores were recorded and entered into columns 1 and 2 of a MINITAB worksheet:

ROW	new	old
1	9	5
2	8	6
3	6	4
4	8	6
5	7	5
6	6	7
7	8	6
8	7	

The instruction to carry out the test is straightforward:

```
MTB > mann-whitney c1 c2
```

First the number of observations (N) and the median for each treatment is given:

```
Mann-Whitney Confidence Interval and Test
new    N =   8   Median =   7.500
old    N =   7   Median =   6.000
```

Now MINITAB calculates an estimate of the difference between the two medians and provides a confidence interval for this difference:

```
Point estimate for ETA1–ETA2 is 2.000
95.7 pct c.i. for ETA1–ETA2 is (1.000, 3.000)
```

This seems rather odd at first sight. ETA1–ETA2 represents the difference between the 'centre' of the new compost scores and the 'centre' of the old compost scores. ETA stands for a Greek letter of that name. If we take away one median from another we have 7.5 – 6 = 1.5. However, the computer uses a more sophisticated calculation to estimate this difference. It calculates the median of all the pairwise differences between the observations in the two treatments. This gives a value of 2.0.

The confidence interval (1.0 – 3.0) is at 95.7%. This is the range of values for which the null hypothesis is not rejected and is calculated for a confidence level as close as possible to 95% (again the method involves repetitive calculations which can only sensibly be carried out by a computer program).

Finally a test statistic, W, is calculated. All the scores are ranked together with the smallest observation given rank 1 and the next largest rank 2, etc. If two or more scores are tied the average rank is given to each. Then the sum of the ranks of the new compost is calculated:

score	4	5	5	6	6	6	6	6	7	7	7	8	8	8	9
compost	O	O	O	O	O	O	N	N	O	N	N	N	N	N	N
rank	1	2.5	2.5	6	6	6	6	6	10	10	10	13	13	13	15

(for example, the two trees with a score of 5 each receive a rank of 2.5 because this is the average of the next two available ranks, 2 and 3)

The sum of the ranks for the new compost trees (the smaller sample size) is:

6 + 6 + 10 + 10 + 13 + 13 + 13 + 15 = 86. This is W:
W = 86.0

Then the p value for the test is given (p = 0.0128):

Test of ETA1 = ETA2 vs. ETA1 n.e. ETA2 is significant at 0.0128

This calculation assumes that there are no tied values in the data. Since there are some ties in this case an adjustment has to be made to the calculation of the probability level to give p = 0.0106:

The test is significant at 0.0106 (adjusted for ties)

We can conclude that the two composts differ in their effect on tree growth form. The new compost is preferable as it produces trees with a higher median score. We now need to consider the relative costs and benefits of using the two composts.

9.3 KRUSKAL-WALLIS TEST

Suppose that we have data from an experiment with more than two treatments. We want to ask questions of it as we would in analysis of variance but we should not because the observations suggest that the populations are not normally distributed. We can then consider using a Kruskal-Wallis test because this test makes less stringent assumptions about the nature of the data than does analysis of variance. However, it still makes the same assumptions as the Mann-Whitney test does: first, the observations must be random and independent observations from the populations of interest; and second, the samples we compare are assumed to come from populations which have a similar shaped distribution. This does not have to be 'normal'. It could be that both/all tend to have a few large values and so have 'positive skewness' (Fig. 9.1a) or they could both/all be negatively skewed but whatever the shape they must both/all share it.

We will illustrate the use of this test on two datasets. First we will re-examine the data we analysed in Chapter 6 using one way analysis of variance.

We should note that data which are suitable for analysis of variance can always be analysed by Kruskal-Wallis but the reverse is not always so. Where the option of using analysis of variance exists however it will normally be preferred since it is a more sensitive and elegant technique (for example, the analysis of a factorial structure of treatments is easily achieved).

To remind you, the data are from four treatments (types of vegetation management by sowing and cutting) each replicated four times on plots whose positions were randomly selected around a field margin. The data are the number of spiders per plot.

```
MTB > print c1 c2

ROW   treat   spider
  1     1       21
  2     1       20
  3     1       19
  4     1       18
  5     2       16
  6     2       16
  7     2       14
  8     2       14
  9     3       18
 10     3       17
 11     3       15
 12     3       16
 13     4       14
 14     4       13
 15     4       13
 16     4       12
```

We ask MINITAB to carry out the Kruskal-Wallis test on the data:

MTB > Kruskal-Wallis c1 c2

LEVEL	NOBS	MEDIAN	AVE. RANK	Z VALUE
1	4	19.50	14.4	2.85
2	4	15.00	7.0	−0.73
3	4	16.50	9.9	0.67
4	4	13.00	2.7	−2.79
OVERALL	16		8.5	

First, the number of observations in each treatment is presented (NOBS), together with their median values. All 16 observations are ranked and the mean rank of each treatment is given. (If there are ties the mean rank is given to each individual.) Finally, a 'z value' is calculated for each treatment. This shows how the mean rank for each treatment differs from the mean rank for all 16 observations and details of its calculation are given in the box below.

For convenience, the mean rank for all observations has been converted to zero. This overall mean has been subtracted from each treatment's mean rank and the result has been divided by the standard deviation of that treatment's ranks to give a z value for each treatment. These show the location of each treatment's mean rank around zero, the overall mean of this 'standard' normal distribution (Fig. 9.2). We remember from Chapter 2 that 95% of the values would be expected to lie between + 1.96 and − 1.96 units from the mean. Here we see that two treatments have z values outside this range suggesting that not all treatments are likely to come from one population.

The program then prints a test statistic (H) (just as we have met t and F before):

H = 12.66   df = 3   p = 0.006

The degrees of freedom are as usual one less than the number of treatments. The p value is given (0.006), so we don't need to consult statistical tables. We can reject the null hypothesis that the observations all come from the same population, because the value of p is very much less than 0.05. We have strong evidence that at least one of the four treatments differs from at least one of the other four in terms of its spider population.

H = 12.85   df = 3   p = 0.005 (adj. for ties)

A second version of the test statistic, H, is also given. This differs slightly

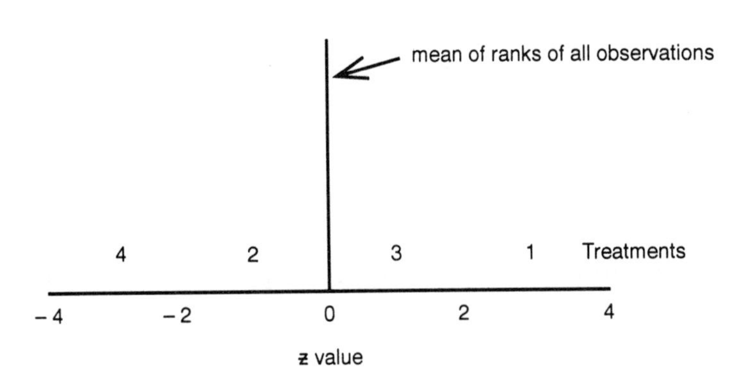

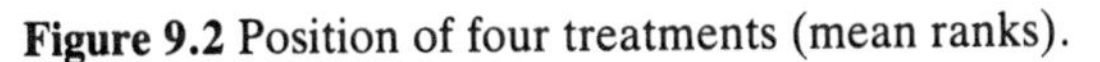

Figure 9.2 Position of four treatments (mean ranks).

from the first in that an adjustment has been made to account for the presence of tied observations (for example there are several plots with 16 spiders). This has to be done because the test assumes that the observations come from a continuous distribution (which could contain values like 2.13459) whereas we have used it to analyse counts where ties are more likely to occur. We should use this second, corrected version of the statistic.

Let's now analyse the results of a different experiment. A student has scored the amount of bacterial growth on each of 20 petri dishes. A score of zero indicates no growth while one of 5 shows that the bacterium covers the entire dish. Five days previously, a randomly selected ten of the dishes had received a standard amount of an established antibiotic used to inhibit bacterial growth (treatment 1) while the remaining ten dishes had received the same amount of a newly discovered inhibitory compound (treatment 2).

MTB > print 'treat' 'score'.

ROW	treat	score
1	1	1
2	1	2
3	1	3
4	1	2
5	1	3
6	1	4
7	1	2
8	1	1
9	1	2
10	1	3
11	2	3
12	2	3
13	2	5

14	2	4
15	2	2
16	2	5
17	2	4
18	2	3
19	2	4
20	2	3

A preliminary plot of the data suggests that the new compound does not seem to be an improvement:

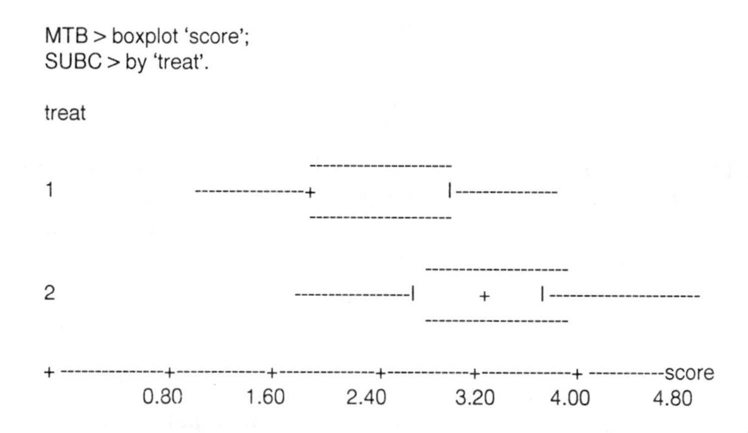

The Kruskal-Wallis test tests the null hypothesis of no difference between the two populations.

```
MTB > Kruskal-Wallis 'score' 'treat'.

LEVEL      NOBS  MEDIAN   AVE.  RANK   Z VALUE
1           10    2.000          7.2    -2.46
2           10    3.500         13.8     2.46
OVERALL     20

H = 6.0   d.f. = 1  p = 0.014
H = 6.46  d.f. = 1  p = 0.011 (adj. for ties)
```

We have evidence to reject the null hypothesis ($p < 0.05$). We conclude that the new chemical is a worse inhibitor of bacterial growth.

9.4 FRIEDMAN'S TEST

In analysis of variance we saw that blocking was a useful way of taking account of some of the random variation and so making it easier to detect the effect of treatments (if there are any such effects). Friedman's test allows us to do the same thing. However it is a nonparametric test and, as with the Kruskal-Wallis test, it allows us to analyse data from populations which are not normally distributed but which have a similar shape.

In the field experiment with sowing and cutting management treatments we allocated one of each treatment to each of four blocks, with a block being one side of the field.

```
MTB > print c1 c2 c3

ROW     treat   spider   block
1       1       21       1
2       1       20       2
3       1       19       3
4       1       18       4
5       2       16       1
6       2       16       2
7       2       14       3
8       2       14       4
9       3       18       1
10      3       17       2
11      3       15       3
12      3       16       4
13      4       14       1
14      4       13       2
15      4       13       3
16      4       12       4

MTB > Friedman c1 c2 c3

Friedman test of spider by treat blocked by block

S = 12.00  d.f. = 3  p = 0.008

                        Est.     Sum of
treat       N         Median     RANKS
1           4         19.344     16.0
2           4         14.844      8.0
3           4         16.469     12.0
4           4         12.719      4.0

Grand median = 15.844
```

As with the Kruskal-Wallis test, the number of observations in each treatment is given together with their median values. Here the sum of the ranks of each observation within each block is then calculated. The sum of 16 for treatment 1 shows that it received the maximum rank of 4 in each of the four blocks. Above this summary lies the test statistic, S, together with the degrees of freedom (4 treatments –1) and the p value (0.008). This is slightly smaller than the p value (0.014) obtained by ignoring blocking (Kruskal-Wallis). In other words, as we might expect in a well-designed experiment blocking has caused an increase (albeit small in this case) in the efficiency of detecting differences between treatments.

Returning to our second example of the effect of two chemicals on bacterial growth we can see that a Friedman's test might also be appropriate here. Suppose that it takes about 30 seconds to score a petri dish for bacterial growth. Those that are last to be scored will have had longer to grow than those scored first. Such a difference could be important if, for example, all the treatment 1 dishes were scored first, then all the treatment 2 ones. It would be much better to arrange our dishes in ten pairs, each containing one of each treatment. We then score each pair, randomly choosing whether to score treatment 1 or treatment 2 first within each pair. We can then take time of assessment into account in the analysis as a blocking effect. The two dishes which were scored first are given a block value of 1 while the last pair are each given a value of 10. This will remove any differences in growth due to time of assessment from the analysis.

MTB > print c1–c3

ROW	treat	score	time
1	1	1	1
2	1	2	2
3	1	3	3
4	1	2	4
5	1	3	5
6	1	4	6
7	1	2	7
8	1	1	8
9	1	2	9
10	1	3	10
11	2	3	1
12	2	3	2
13	2	5	3
14	2	4	4
15	2	2	5
16	2	5	6
17	2	4	7
18	2	3	8
19	2	4	9
20	2	3	10

MTB > Friedman 'score' 'treat' 'time'.
Friedman test of score by treat blocked by time

S = 4.90 d.f. = 1 p = 0.027
S = 5.44 d.f. = 1 p = 0.020 (adjusted for ties)

treat	N	Est. Median	Sum of RANKS
1	10	2.000	11.5
2	10	4.000	18.5

Grand median = 3.000

Here the test statistic, S (as for the H statistic in the Kruskal-Wallis test), is still highly significant ($p < 0.05$), but the p value has not been reduced compared with that from the Kruskal-Wallis test. It appears that time of assessment is not very important here. After all, with only 20 plates in total the assessment should be complete in 10 minutes. However, if the bacteria grew very fast and if there were 100 plates to assess, taking at least 100 minutes, then blocking for time could be very important.

9.5 CHI-SQUARED CONTINGENCY TEST

Up to now the results of our experiments have provided us with an observation for each individual (plot or petri dish) which is **quantitative**. They may be continuous data (weights or lengths or speeds) or counts (how many spiders) or scores (the amount of bacterial growth on a 0–5 scale).

In contrast we may have an experiment in which each observation for each treatment is **categorical**. This means that it can be placed in one of two or more categories, for example: green or brown; live or dead; small, medium or large. If we prepare a table with two or more rows and two or more columns to summarize the data each item in the table will represent the number of observations in that category. We then want to ask questions like: is the proportion of green individuals the same in each treatment? Let us clarify this with an example.

A new drug has been developed which may be more or less effective at clearing all parasites from the blood of humans within 36 hours. In an experiment 287 individuals took part to compare its effectiveness with that of chloroquinine (the standard). Of the 184 individuals receiving chloroquinine, 129 were cleared of parasites within 36 hours while 55 were not. We can summarize these observations and those for the new drug in a table of **observed** values (0):

	Cleared in 36 hr	Not cleared in 36 hr	Total
Chloroquinine	129	55	184
New drug	80	23	103
Total	209	78	287

Note that the numbers of individuals taking the new drug and those taking the old one do not have to be equal although the analysis will be more robust if they are of similar magnitude. Our null hypothesis is that the two variables are statistically independent: the proportion of individuals from whose blood the parasite has been cleared is the same for both drugs. Because we have only a sample from each of the two populations we need statistically to assess these data. The first step is to calculate the number of individuals we would expect to be in each category (or 'class' or 'cell') if the null hypothesis is true. For the top left class of the above dataset (Chloroquinine, Cleared) this is:

$$\text{Expected value} = \text{E} = \frac{209 \times 184}{287} = 133.99$$

In general:

$$\text{Expected value} = \frac{\text{row total} \times \text{column total}}{\text{grand total}}$$

The maximum number of people who could appear in this category is 184 because this is the number who were treated with chloroquinine. The proportion 209/287 is the overall proportion of individuals from whose blood the parasite had cleared in 36 hours, irrespective of which drug they had received. This proportion is then multiplied by the number of individuals who received chloroquinine (184) to give the number of individuals we would expect to find who had both received chloroquinine and whose blood had cleared in 36 hours assuming both drugs have the same effect. This value represents H_0.

This test is called a chi-squared **contingency** test because the expected values are contingent upon (dependent upon) the row and column totals and the grand total. The table of such **expected** values (E) is then:

	Cleared in 36 hr	Not cleared in 36 hr	Total
Chloroquinine	133.99	50.01	184
New drug	75.01	27.99	103
Total	209	78	287

The general idea is that if the null hypothesis is true, the observed and expected counts will be very similar whereas if it is false the counts will be very different. To enable us to calculate the strength of evidence against the null hypothesis we calculate a test statistic – 'chi-squared' (chi is a Greek letter pronounced 'ky' to rhyme with sky):

$$\chi^2 = \sum \frac{(O-E)^2}{E}$$

χ^2 is calculated as the sum of (the squares of the differences between each observed count and its expected value, divided by its expected value). Here we have:

$$(129 - 133.99)^2/(133.99) +$$
$$(55 - 50.01)^2/(50.01) +$$
$$(80 - 75.01)^2/(75.01) +$$
$$(23 - 27.99)^2/(27.99) = 1.91$$

The bigger the value of χ^2 the greater the chance that we will reject the null hypothesis. Here, $\chi^2 = 1.91$. We compare this with a critical value from a statistical table of chi-square taken from the column headed 95% ($p = 0.05$). If 1.91 is bigger than the critical value then the null hypothesis is rejected with 95% confidence. As always we need to know the degrees of freedom to find the critical value in statistical tables. In a contingency table there are (r-1)x(c-1) degrees of freedom where r = the number of rows and c = the number of columns. Thus for a 5 × 4 table the df = 12. Here the df = 1 and the calculated value of $\chi^2 = 1.91$ is less than the table value of 3.841 so we have no reason to reject the null hypothesis of no difference in the efficacy of the two drugs.

MINITAB can carry out a chi-squared analysis for us. The contents of the table of observed counts is placed into the worksheet:

```
MTB > print c11 c12

ROW   clear   not
  1     129    55
  2      80    23
```

```
MTB > chisquare c11 c12

Expected counts are printed below observed counts

            clear       not      Total
1             129        55        184
           133.99     50.01

2              80        23        103
            75.01     27.99

Total         209        78        287

ChiSq= 0.186+ 0.499+
0.332+ 0.891= 1.908

df = 1
```

MINITAB prints the expected values in each category below the observed values and presents the row and column totals and grand total. Then follow the components of chi-squared (that is the value of 'observed–expected', squared and divided by expected) which are added up to give the chi-squared statistic, together with its degrees of freedom.

Here the chi-squared statistic is very small, indicating that the observed values are very close to those we expected to find if the null hypothesis were true. Formally, we should compare our value of 1.908 with the value of χ^2 found in a statistical table for 1 df and $p = 0.05$. This is 3.84. Since 1.908 is less than 3.84 we conclude that a similar proportion of people is cured by the two drugs.

9.5.1 Some important requirements for carrying out a valid chi-squared contingency test

1. The observations must be counts not percentages or proportions or measures. It is important that the grand total is the number of independent individuals in the study as this gives information about the reliability of the estimates.

2. In general the **expected** count or frequency in each class should exceed 2 and also 80% of the classes should have expected frequencies greater than 5. If this is not so then either we should collect more data or we can combine neighbouring classes (if this is sensible).

For example, if we applied fertilizer to 30 of 60 plants at random and classified the plants after 1 month's growth we might find:

	small	medium	large
no fertilizer	20	7	3
with fertilizer	10	15	5

The expected value for the top right class (large, without fertilizer) would be (30 × 8)/60 = 4. For the bottom right class (large, with fertilizer) would be the same. To fulfil the requirements for a chi-squared contingency test we can combine the medium and large categories:

	small	large
no fertilizer	20	10
with fertilizer	10	20

Now there will be no problem with expected values being too small. They all happen to be (30 × 30)/60 = 15.

The reason for ensuring that expected values are greater than 5 is that the test is over-sensitive to small differences when the expected value is small. This is because dividing by a very small expected value (imagine dividing by 0.1) will give rise to a ridiculously high component of chi-squared.

3. When the contingency table is larger than the 2 × 2 case look at the component of χ^2 which comes from each class. Those classes with large values are mainly responsible for the rejection of the null hypothesis and are therefore the ones to concentrate on when it comes to interpreting your results, as we can see in the following example.

9.5.2 A further example of a chi-squared contingency test

We have carried out an experiment to compare the effectiveness of three termite repellents in preserving fencing stakes. Each chemical was applied to a sample of 300 fencing stakes, giving 900 stakes in total. The number of stakes which were attacked was recorded after a year in the ground.

```
MTB > print c1 c2

ROW   attack   avoid
  1     112     188    oil
  2      82     218    creosote
  3     123     177    copper arsenate
```

```
MTB > chis c1 c2
Expected counts are printed below observed counts

          attack      avoid      Total
1           112         188        300
          105.67      194.33

2            82         218        300
          105.67      194.33

3           123         177        300
          105.67      194.33

Total       317         583        900

ChiSq = 0.380 + 0.206 +
        5.301 + 2.882 +
        2.843 + 1.546 = 13.158

df = 2
```

The table value of chi-squared for 2 df at p = 0.05 is 5.99 and for p = 0.01 it is 9.21. We therefore have strong evidence to reject the null hypothesis that all three chemicals are equally effective.

If we examine the components of chi-squared we see that the two coming from creosote are especially high (5.301 and 2.882). This tells us that it is creosote which has a different effect from the others. Comparing the observed and expected values we see that creosote was more effective than the other two treatments with only 82/300 = 27% of stakes being attacked compared with 37% and 41% for the others.

An important assumption we have made is that the allocation of treatments to the 900 stakes was carried out at random. If the 300 stakes for each treatment had been grouped any differences between treatment might have been due to environmental differences such as one treatment being on a sandier soil which is better shaded, thus possibly affecting the density of termites in the area. As with many other forms of sampling, therefore randomization is the key.

10 Summarizing data from an observational study

In an experiment we must have an observation from each of several **experimental units** from each treatment (**replication**). These replicates must not have any effect on each other (**independence**) and they should be arranged during the experiment so that each one has an equal chance of being in any particular position (**randomization**). However, the name 'experiment' is often mistakenly applied to investigations which are really observational studies. For example, the investigator makes observations without applying any treatments or in which the so-called replicate plots of each treatment are grouped together, i.e. they are neither randomized nor interspersed (Fig. 10.1).

In this chapter we will outline an exploratory technique which can simplify the results from such observational studies – **principal components analysis**. Its great strength is that it can summarize information about many different characteristics recorded from each individual in the study. We will illustrate its use by reference to two datasets: first, a study of the

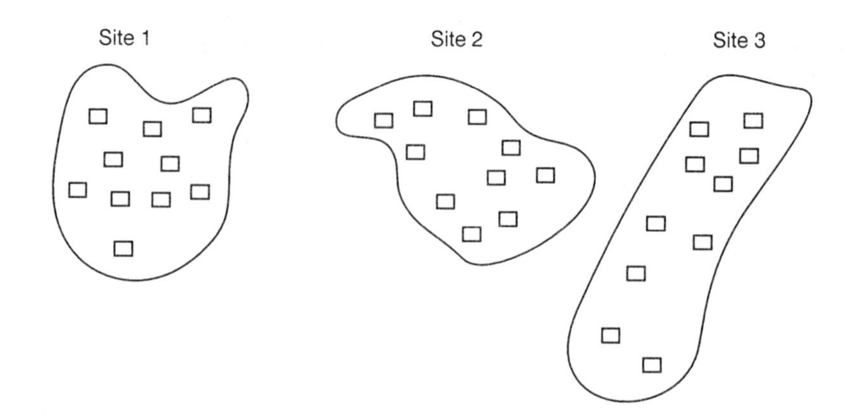

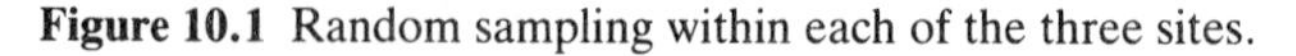

Figure 10.1 Random sampling within each of the three sites.

economic and social characteristics of a group of countries and second, a study of the vegetation on a hillside.

10.1 CASE STUDY 1. ASSESSING THE ECONOMIC AND SOCIAL CHARACTERISTICS OF 14 COUNTRIES

We have obtained information about seven characteristics (population density, proportion in agricultural employment, national income per head, investment in machinery, etc., infant mortality rate, energy consumption per head and the number of TV sets per hundred of the population) for 14 countries (Australia, France, West Germany, Greece, Iceland, Italy, Japan, New Zealand, Portugal, Spain, Sweden, Turkey, the UK and the USA) (O.E.C.D. Observer, **109**, March 1981). This has been entered on MINITAB in columns 1 to 7 in the same order as listed above. Each column has been given a suitable short name (from 1 to 8 characters).

MTB > print c1–c7

ROW	popden	agemp	natinc	capinv	infmort	energy	tvsets
1	2	6	8.4	10.1	12	5.2	36
2	97	9	10.7	9.2	10	3.7	28
3	247	6	12.4	9.1	15	4.6	33
4	72	31	4.1	8.1	19	1.7	12
5	2	13	11.0	6.6	11	5.8	25
6	189	15	5.7	7.9	15	2.5	22
7	311	11	8.7	10.9	8	3.3	24
8	12	10	6.8	8.0	14	3.4	26
9	107	31	2.1	5.5	39	1.1	9
10	74	19	5.3	6.9	15	2.0	21
11	18	6	12.8	7.2	7	6.3	37
12	56	61	1.6	8.8	153	0.7	5
13	229	3	7.2	9.3	13	3.9	39
14	24	4	10.6	7.3	13	8.7	62

We then obtain **summary statistics** for each of the variables (see p. 21 for explanation of the columns):

MTB > describe c1–c7

	N	MEAN	MEDIAN	TRMEAN	STDEV	SEMEAN
popden	14	102.9	73.0	93.9	101.3	27.1
agemp	14	16.07	10.50	13.42	15.73	4.20
natinc	14	7.671	7.800	7.750	3.619	0.967
capinv	14	8.207	8.050	8.208	1.462	0.391
infmort	14	24.6	13.5	15.3	37.7	10.1
energy	14	3.779	3.550	3.625	2.211	0.591
tvsets	14	27.07	25.50	26.00	14.40	3.85

	MIN	MAX	Q1	Q3
popden	2.0	311.0	16.5	199.0
agemp	3.00	61.00	6.00	22.00
natinc	1.600	12.800	5.000	10.775
capinv	5.500	10.900	7.125	9.225
infmort	7.0	153.0	10.8	16.0
energy	0.700	8.700	1.925	5.350
tvsets	5.00	62.00	18.75	36.25

The MINITAB printout shows that there is a great deal of variability between countries, but the data are not easy to digest. If we had only obtained information about one variable, such as population density, we might draw a boxplot. Even with seven variables it is sensible to start by looking at each in this way, to spot outliers:

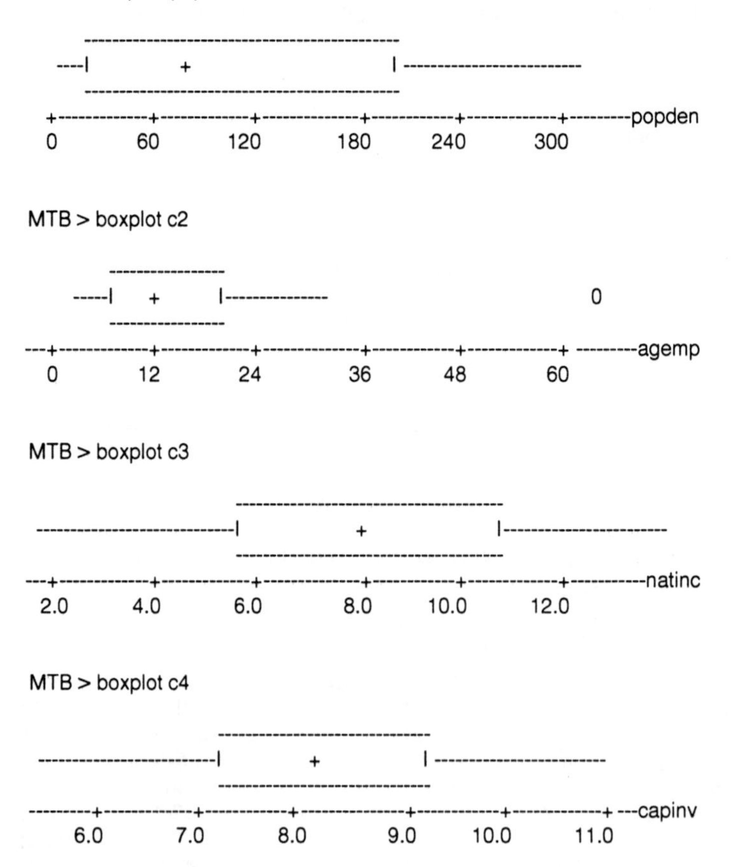

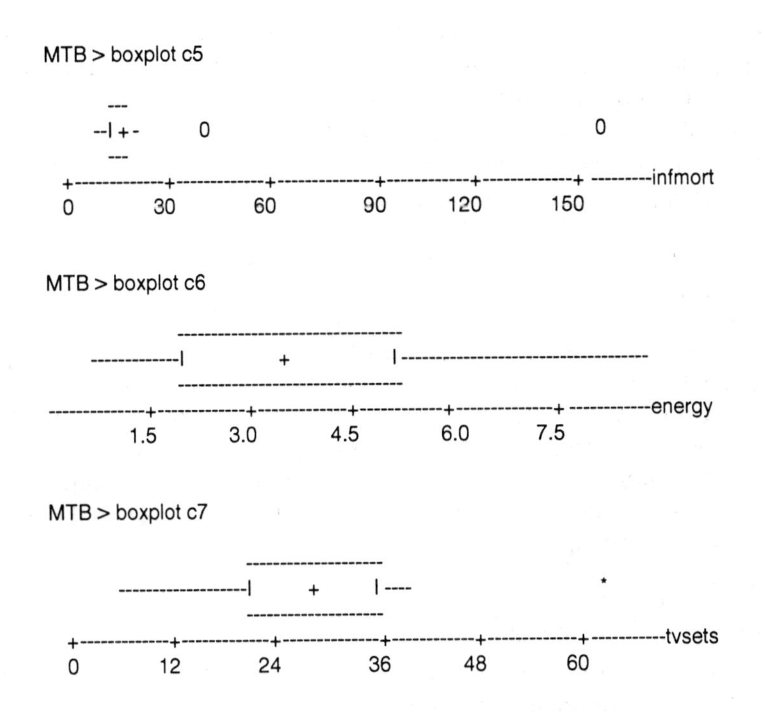

We see that Turkey had an unusually high proportion of agricultural employment and that both it and, to a much lesser degree, Portugal had a high infant mortality rate, while the USA had a high number of TV sets per hundred people.

Some of these variables are probably correlated. We can check for this by asking for correlation coefficients (Chapter 8):

MTB > corr c1–c7

	popden	agemp	natinc	capinv	infmort	energy
agemp	–0.150					
natinc	0.019	–0.786				
capinv	0.490	–0.183	0.196			
infmort	–0.131	0.890	–0.602	0.002		
energy	–0.255	–0.715	0.830	0.009	–0.494	
tvsets	–0.069	–0.783	0.722	0.134	–0.526	0.915

Here we see, for example, a strong positive correlation between energy and TV sets, while there is a strong negative correlation between the proportion of the population in agricultural employment and TV sets.

It would be useful to be able to give each country a score which took into account all seven of its characteristics, then these scores would summarize all the relative similarities and differences between them. We might consider simply adding up the seven values for each country. However, this would give undue weight to variables with large values (like population density). So we could start by **standardizing** the data. If we subtract the mean population density from each individual value and then divide by the standard deviation of population density the new values have a mean of zero and a variance of one. They have been standardized. The same method is applied to the remaining six variables so that they all now have means of zero and variances of one. In this way they all have equal status.

We can, however, still improve upon this by adding together these new values for each variable for each country to obtain a score. We can ask for our scores to be combined in such a way as to explain as much of the variation present in our dataset as possible. This is the basis for principal components analysis (pca). It provides us with the 'best' scores.

We ask for a pca to be carried out on the seven variables in columns 1 to 7, with the 'best' scores for the first principal component to be put in column 11 (with a further set of scores which accounts for some of the remaining variation being put into column 12 – the second principal component and so on).

```
MTB > pca c1–c7;
SUBC > scores c11–c17.
```

The resulting printout can seem rather overwhelming, but we often need only concentrate on small parts of it.

Eigenanalysis of the Correlation Matrix

Eigenvalue	3.9367	1.5639	0.8100	0.3572	0.2703	0.0451
Proportion	0.562	0.223	0.116	0.051	0.039	0.006
Cumulative	0.562	0.786	0.902	0.953	0.991	0.998

Eigenvalue	0.0169
Proportion	0.002
Cumulative	1.000

The first part of the output (above) has seven columns. Each represents one of seven principal components which summarize the variability in the data and each contains an **eigenvalue**. We can think of the eigenvalue as the amount of variation in the dataset which is explained by that particular principal component. For example the first principal component has an eigenvalue of 3.9367. This represents 56.2% (shown underneath the eigenvalue in row two, but as a proportion – 0.562) of the sum of all seven eigenvalues. Principal component number two accounts for a further 22.3%, so components 1 and 2 account for 78.6% (given in row three as 0.786) of all the variation between the two of them.

This is good news; we have accounted for over three-quarters of the variation in the data by two sets of scores whereas the original data contained seven variables. If we include the third principal component we can account for 90% of the variation. However we shall probably obtain enough insight into the structure of the data by examining only the first two principal components.

The printout then shows the coefficient or weight allocated to each variable in each of the seven principal components to account for the maximum amount of variation and to produce the 'best scores':

Variable	PC1	PC2	PC3	PC4	PC5	PC6
popden	–0.009	0.717	0.252	0.634	0.060	–0.116
agemp	0.476	–0.108	–0.257	0.156	0.158	–0.598
natinc	–0.452	0.015	–0.146	0.022	0.805	0.189
capinv	–0.083	0.639	–0.538	–0.519	–0.116	–0.114
infmort	0.394	–0.067	–0.626	0.379	0.097	0.462
energy	–0.450	–0.234	–0.311	0.227	0.016	–0.577
tvsets	–0.452	–0.085	–0.267	0.329	–0.549	0.182

Variable	PC7
popden	–0.059
agemp	0.537
natinc	0.301
capinv	0.003
infmort	–0.286
energy	–0.511
tvsets	0.524

What are these coefficients? How are they used? Let us take the column headed PC1. We will use these coefficients to obtain the score for the first country (Australia) on principal component 1. To do this we will copy down the coefficients and then put the actual data for Australia's population density, etc., beside them (check that you can see that the data come from row 1 in the printout at the beginning of this section):

Variable	PC1	Data
popden	–0.009	2
agemp	0.476	6
natinc	–0.452	8.4
capinv	–0.083	10.1
infmort	0.394	12
energy	–0.450	5.2
tvsets	–0.452	36

We now standardize each data value. This is achieved by subtracting the mean of that variable and then dividing by its standard deviation. For example, for population density we subtract 102.9 and then divide by 101.3. For Australia this gives –0.996. The standardized values for the other variables are given below. Check that you can obtain the standardized value for agricultural employment in the same way by referring to the data summary on p. 139 to find the mean and standard deviation.

Variable	PC1	Data	Standardized Data
popden	–0.009	2	–0.996
agemp	0.476	6	–0.640
natinc	–0.452	8.4	0.201
capinv	–0.083	10.1	1.295
infmort	0.394	12	–0.334
energy	–0.450	5.2	0.643
tvsets	–0.452	36	0.620

Now we multiply the coefficient and the standardized value for each variable to give a final column of numbers which, when added together, give the score for Australia.

Variable	PC1	Standard Data	PC1 × Standard Data
popden	–0.009	–0.996	0.008964
agemp	0.476	–0.640	–0.304640
natinc	–0.452	0.201	–0.090852
capinv	–0.083	1.295	–0.107485
infmort	0.394	–0.334	–0.131596
energy	–0.450	0.643	–0.289350
tvsets	–0.452	0.620	–0.280240
SUM = SCORE on PC1 axis			–1.195199

We ask MINITAB to print out this score and those for the other 13 countries for the first two principal components whose columns we have named pc1 and pc2. Notice that our value for the score of Australia on PC1 (–1.195199) agrees with the MINITAB value (–1.19523) to three figures after the decimal point.

```
MTB > print c11–c12

ROW      pc1        pc2

  1   –1.19523    0.00515
  2   –0.81330    0.48198
  3   –1.41233    1.39261
  4    1.74510   –0.06283
  5   –0.89650   –1.55911
  6    0.54333    0.65681
  7   –0.43256    2.78599
  8   –0.05447   –0.62952
  9    2.56455   –0.91376
 10    0.91468   –0.56365
 11   –1.88907   –1.24519
 12    4.74817   –0.17634
 13   –0.93008    1.39480
 14   –2.89230   –1.56695
```

We now put the code numbers 1 to 14 for each country into column 8.

```
MTB > set c8
DATA > 1:14
DATA > end
```

We ask for a labelled plot of pc2 against pc1 with each country being labelled according to its code number in column 8 (MINITAB will replace the code numbers 1 to 14 by the first 14 letters of the alphabet (A – N)).

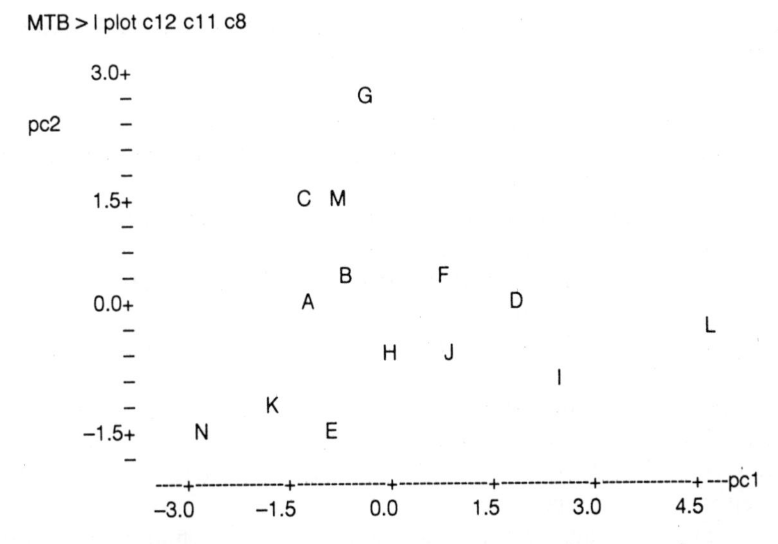

This graph summarizes our data. We see that the countries are well spread out. Let us take the first principal component (x axis). What aspects of the data does it represent? The weights or coefficients for agricultural employment and infant mortality are large and positive. This means that countries with a high proportion of people in agricultural employment and with high infant mortality (Turkey) are at the right-hand side of the graph. The weights for national income, energy and TV sets are large and negative. This means that countries with high national income, high energy consumption and a large number of TV sets per 100 people are at the left-hand side of the graph (Sweden and the USA). With this information we can reasonably summarize the meaning of pc1 as a summary statistic of economic development on which a low score represents a high level of development.

Looking at pc2 we see that countries at the top of the graph are characterized by a high population density and a high level of capital investment (Japan) whereas those at the bottom of the graph have a low population density and a lower amount of capital investment (Iceland, USA). This graph enables us to see the similarities and differences between the 14 countries at a glance. In this sense it is a great improvement on the original table of data. We could if we wanted continue with the other components. For example, we would next look at component 3 (PC3).

However it is not very worthwhile because it only accounts for a further 11% of the variation and the remaining components only account for about the same amount of variation again between them.

10.2 CASE STUDY 2: SPECIES COMPOSITION OF GRASSLAND ON THREE PARTS OF A SITE

A common exercise on many ecology courses is to put ten 0.5 m × 0.5 m quadrats down at random on the top of a hill, another ten on one face and another ten at the bottom of the hill. We then estimate the percentage cover of each species in each quadrat. The aim is to find out whether and, if so, how the vegetation differs between the areas.

It is tempting to see this as an experiment with three treatments (sites) and ten replicates but really it is separate random sampling within each of three separate areas (Fig. 10.1). The results can be summarized separately for each area; for example by giving estimates of the mean number of species per quadrat with a standard deviation to show the variability within the site (Fig. 10.2). However, what we should not do is use the method of analysis of variance to answer the question 'Is there evidence of a difference in percentage cover of a particular species between the three areas?'. This is because we have not started from a common population and then applied the treatments (top, middle, bottom) at random.

Instead we should use an exploratory technique, such as principal components analysis, to summarize the variability. This has the added

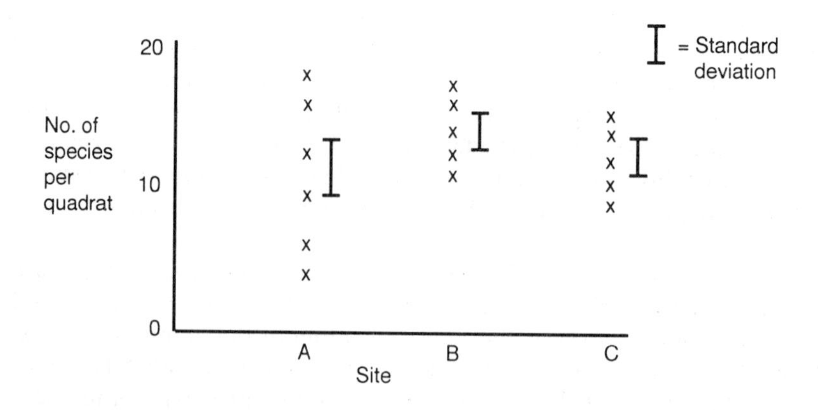

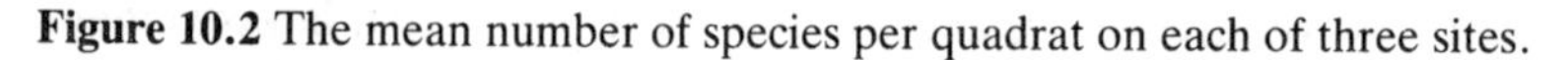
Figure 10.2 The mean number of species per quadrat on each of three sites.

advantage of using the information about all the species together. It is a type of **multivariate** analysis.

We enter the data into MINITAB, with 37 columns (= species) and 30 rows (= quadrats).

Here are the data (collected by Wye College students) for the first six species = columns which represent *Agrostis capillaris*, *Brachypodium pinnatum*, *Carex caryophyllea*, *Carex flacca*, *Dactylis glomerata* and *Festuca ovina*.

0	40	0	0	53	100
0	66	0	0	100	100
0	70	0	0	70	90
0	53	0	0	47	87
17	83	0	0	22	89
13	87	0	0	87	93
100	100	0	0	80	90
50	0	0	0	40	80
0	0	0	0	0	100
20	50	0	0	60	90
0	90	40	100	0	100
0	40	40	60	0	60
0	50	25	88	0	100
0	88	38	63	0	88
0	50	10	30	5	65
0	100	20	50	50	100
0	10	80	100	0	90
0	60	40	80	0	70
0	80	0	20	0	0
0	60	0	0	40	90
0	0	0	0	60	40
0	0	0	0	80	80
40	0	0	0	40	40
60	0	0	0	0	0
0	8	0	0	0	100
20	20	0	20	40	0
40	0	0	0	20	100
0	0	0	0	0	0
40	0	0	0	40	100
40	50	0	0	50	90

We ask for a principal components analysis:

```
MTB > pca c1–c37;
SUBC > scores c40 c41.
```

and name column 40 'pc1' and column 42 'pc2'.

The output below shows the proportion of variation accounted for by each of the first six principal components (eigenvalues). (I have removed the seventh principal component from the copy of the printout to save space.) Then the weights or coefficients for each species in each of the first six principal components have been displayed.

Eigenanalysis of the Correlation Matrix

Eigenvalue	11.544	5.144	2.639	2.257	2.016	1.773
Proportion	0.312	0.139	0.071	0.061	0.054	0.048
Cumulative	0.312	0.451	0.522	0.583	0.638	0.686

Here the first two axes only account for 45% of the total variation but, as we shall see, it will still provide a useful summary.

Variable	PC1	PC2	PC3	PC4	PC5	PC6
ag.cap*	−0.121	−0.032	−0.248	−0.152	0.277	−0.257
brac.pin	0.120	0.283	−0.114	−0.086	−0.034	−0.024
car.car	0.239	−0.005	−0.060	0.063	0.011	0.046
car.fla	0.268	−0.022	−0.049	0.077	−0.051	−0.015
dac.glo	−0.173	0.190	−0.171	−0.095	−0.156	0.172
fest.ov	0.040	0.196	−0.089	−0.079	−0.259	−0.307
hol.lan	−0.096	0.207	−0.299	−0.173	0.108	0.054
lol.per	−0.177	−0.185	−0.062	0.085	0.019	0.193
ach.mill	−0.080	0.175	0.293	0.219	0.147	−0.050
agr.eup	−0.085	0.173	−0.349	−0.037	−0.219	0.121
cam.rot	0.163	−0.023	0.124	−0.198	−0.282	−0.110
cir.ac	0.183	0.057	0.021	0.171	0.228	−0.005
cir.ar	−0.088	0.219	0.189	0.277	−0.047	−0.102
cir.pal	−0.062	0.203	−0.339	0.022	−0.130	−0.055
crep.spp	−0.072	−0.231	−0.090	0.127	−0.023	0.093
cru.lae	−0.037	0.243	0.137	0.130	0.088	−0.008
cyn.cri	−0.080	−0.197	0.116	0.010	−0.229	0.023
gal.ver	0.012	−0.143	−0.078	0.282	−0.247	0.375
gen.am	0.176	−0.095	−0.194	0.263	0.260	0.025
gle.hed	−0.116	0.247	0.031	0.173	0.155	−0.176
hel.num	0.273	0.017	0.069	−0.092	−0.003	0.134
hie.pil	0.242	−0.003	−0.044	0.145	0.059	0.095
hyp.per	0.172	−0.029	−0.197	0.286	−0.130	−0.112
leon.sp	0.216	−0.074	−0.107	0.005	0.003	−0.112
lot.cor	−0.069	−0.324	−0.017	−0.033	−0.093	−0.315
ori.vul	0.147	0.072	0.113	−0.388	0.061	0.209
pim.sax	0.250	−0.006	−0.025	0.039	−0.137	−0.078
plan.lan	−0.168	−0.313	0.044	0.073	−0.084	0.070
pol.vul	0.135	−0.144	−0.233	0.258	0.114	−0.093
pot.rep	−0.146	0.164	−0.189	0.204	−0.364	0.136
pru.vul	−0.025	−0.182	−0.117	−0.090	−0.147	−0.501
ran.spp	−0.069	0.002	−0.310	−0.014	0.368	0.061
san.min	0.281	0.021	0.046	−0.064	−0.056	0.063
thy.spp	0.254	0.021	−0.043	−0.020	−0.071	0.053
tri.rep	−0.224	−0.193	−0.101	0.020	0.042	0.024
ver.cha	−0.087	0.192	0.167	0.327	−0.100	−0.203
vio.hir	0.232	−0.027	−0.124	0.054	−0.046	−0.082

Each column (species) was given an abbreviated Latin species name (maximum of eight letters allowed).*

We ask for a printout of the scores of each quadrat on the first two principal components from columns 40 and 41 (which we have named appropriately):

```
MTB > print c40 c41

ROW        pc1        pc2
  1   -2.65742    3.23235
  2   -3.11976    3.64558
  3   -2.50492    2.44219
  4   -2.57511    3.52153
  5   -1.33581    2.05404
  6   -0.97591    2.16692
  7   -2.70955    3.06575
  8   -2.10436    1.72044
  9   -1.74847    0.68887
 10   -1.88399    1.47517
 11    6.87997    0.59776
 12    3.72506    1.42079
 13    6.68716   -1.77234
 14    5.48874   -0.24516
 15    2.73888   -0.02376
 16    4.11199    1.25905
 17    6.12395   -0.83651
 18    3.77876   -0.01016
 19    3.09141   -0.28989
 20    2.23702    0.76591
 21   -2.96016   -1.76356
 22   -3.19158   -3.29659
 23   -2.29824   -1.86741
 24   -2.34998   -4.23661
 25   -1.54096   -3.42761
 26   -2.96905   -2.81131
 27   -2.46572   -3.48355
 28   -1.66073   -2.09768
 29   -1.67677   -1.73013
 30   -2.13446   -0.16408
```

We insert the code numbers 1 to 30 in column 39 so that we can then obtain a labelled plot using the corresponding letters of the alphabet:

* Clapham, A.R., Tutin, T.G. and Warburg, E.F. (1981) *Excursion Flora of the British Isles*, Cambridge, CUP.

MTB > set c39
DATA > 1:30
DATA > end

MTB > l plot c41 c40 c39

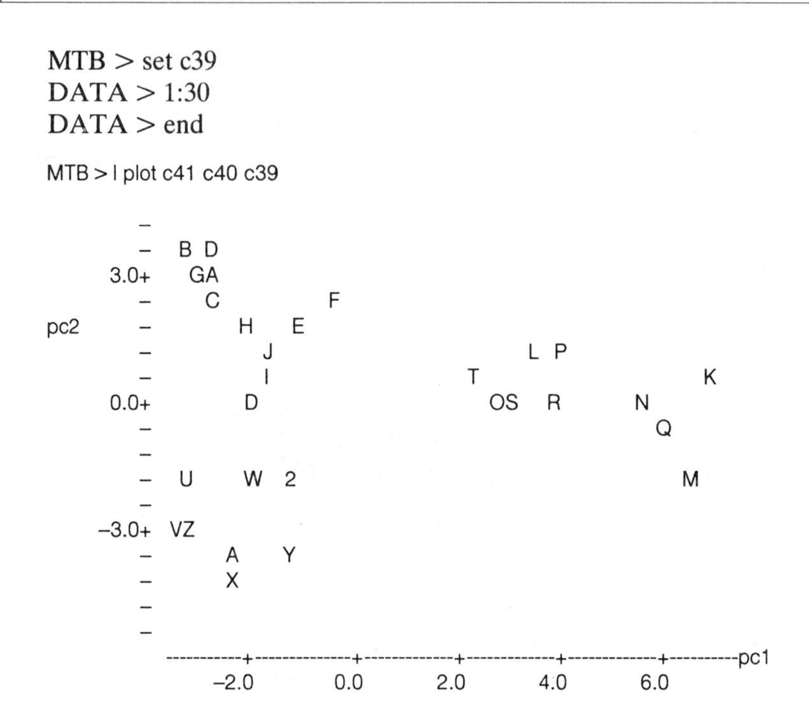

We notice that there are two letter As and two letter Ds on the graph. Because we have 30 quadrats but there are only 26 letters in the alphabet we expect two of A, B, C and D (the figure 2, indicating that two quadrats are very close together must represent a B and a C). We can use the scores for each quadrat to work out that the letters A–D in the lower part of the graph represent the quadrat numbers 26–30.

We can now interpret the meaning of the two axes. For each of the first two components I have selected species with relatively large (positive or negative) coefficients and with whose ecology I am familiar.

Principal component 1

large weights or coefficients (positive and negative)	short name	full names
−0.224	tri.rep.	*Trifolium repens* (white clover)
−0.177	lol.per.	*Lolium perenne* (perennial ryegrass)
−0.168	plan.lat	*Plantago lanceolata* (ribwort plantain)
0.281	san.min.	*Sanguisorba minor* (salad burnet)
0.273	hel.num.	*Helianthemum nummularia* (rock rose)

Principal component 2

large weights or coefficients (positive or negative)	short name	full names
–0.324	lot.cor.	*Lotus corniculatus* (birdsfoot trefoil)
–0.313	plan.lat.	*Plantago lanceolata* (ribwort plantain)
0.283	brac.pin.	*Brachypodium pinnatum* (tor grass)
0.247	gle.hed.	*Glechoma hederacea* (ground ivy)

By referring to textbooks which tell us about the conditions in which these species thrive we can build up a picture of the two axes. The horizontal axis (PC1) goes from a fertile and moist environment on the left (characterized by ryegrass, clover and plantain) to an infertile and dry environment on the right (characterized by salad burnet and rock rose). The vertical axis (PC2) goes from a heavily grazed environment (negative scores) (characterized by birdsfoot trefoil and plantain) to a less-heavily grazed environment (positive scores) (characterized by tor grass and ground ivy).

The quadrats appear to form three groups: numbers 1–10 are at the top left, numbers 11 to 20 at the middle right and numbers 21 to 30 at the bottom left. We can now characterize the areas from which we sampled the vegetation. In the graph below I have enclosed the quadrats from each area:

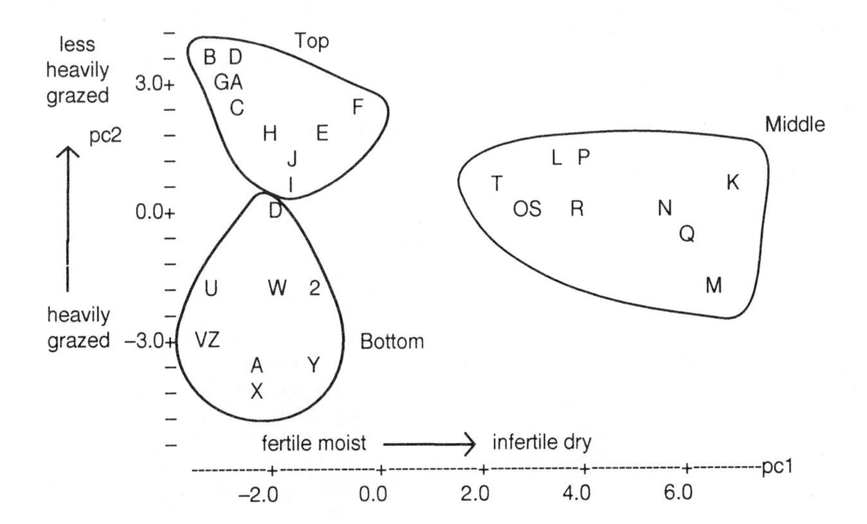

We can conclude that at the top of the hill there is a relatively moist and fertile area but that it is not heavily grazed. In the middle of the slope there is an infertile, dry area which is moderately grazed. At the bottom of the slope is an area which is fertile and moist like the top of the hill but, in contrast, it is heavily grazed.

This type of interpretation has proved to be a very helpful way of summarizing the complex data (a table of $30 \times 37 = 1110$ percentage cover values) which we have obtained.

Your project

11

Most undergraduate courses include a project. This chapter will give you some useful guidelines to help you complete your project successfully.

11.1 CHOOSING A TOPIC AND A SUPERVISOR

If you are short of ideas for a project ask a lecturer for help. Perhaps the lecturer has an idea for a project which may appeal to you. Find out if last year's project reports are available for you to consult, to gain an idea of the sort of topics which might be suitable. There are also books which contain project suggestions (Appendix B).

Start thinking about your project early. You need to find an appropriate supervisor who is willing and able to advise you. There may be only one person who has the experience to supervise a project in the particular subject area which interests you. However, if there is a choice, ideally you should find someone with whom you believe you will get on with and who will also be available to give you feedback at critical times.

It is important to realize that your potential supervisor will probably be a very busy person and you should always arrange to see them by appointment and have a well-prepared list of points ready to discuss. A supervisor will have his or her own other deadlines for teaching, research, administration or pastoral care. He or she will be pleased to help you but you should each be aware of the other's preoccupations because this will help you to plan your joint use of time efficiently.

With this background in mind it is useful to find out whether your potential supervisor will be available to give advice at intervals throughout your project. Will you be able to meet for regular discussion? If not, is there someone else who can help you if you need advice when your supervisor is unavoidably absent?

Pay attention to any deadline for agreeing a project title and outline with your supervisor.

11.2 WHAT CAN HAPPEN IF YOU DO NOT SEEK ADVICE AT THE START

Your supervisor will usually have sufficient knowledge of the nature of your study to be able to recommend a satisfactory method of obtaining and analysing data. However, if there are any doubts he or she may suggest that you should clarify your proposed methodology with a statistician.

It is important to feel confident about your approach to the study. You may be put off seeking advice at the beginning of your project either because you have only just completed a first year statistics course or, more likely, because you have forgotten the statistical principles you covered in your first year course. The subject is still 'unreal' because you do not have your own data in front of you. You may be afraid to reveal your own lack of knowledge of statistics and experimental design or you may be overwhelmed by the sight of mathematical symbols in textbooks. You may even feel that deciding on your objectives and treatments is quite enough of a problem and it is the last straw to discuss the design.

The result of this approach is that you will collect a mass of data and then have to face the awkward question: 'What can I do with these results?'

There are four main ways in which the conversation with your project supervisor will then develop:

At best. A satisfactory analysis is possible which yields reasonable answers to the questions asked. (Unfortunately this is rarely the case.)
Second best. All is well but a more efficient design could have been used and saved a great deal of work.
Third best. Analysis is possible but the experiment was not sensitive enough to answer the question with sufficient precision. Perhaps more replicates were required or extra measurements should have been taken.
Worst. The project is a disaster because the experimental design was badly flawed. Either no inferences can be made, or the results are not relevant to the required questions. In other words, your work was a waste of time and your project is a failure.

It is part of human nature to worry about what you don't know and to fear that other people will ridicule your lack of knowledge. Think carefully about what is concerning you, write down some key points, and make an appointment to ask advice as soon as possible. It's always a great relief to share a concern and to be able to plan ahead with confidence.

Remember, biological material is variable. We need to draw conclusions from experiments or surveys by **induction** from a sample to the whole population. Statistical theory allows definite statements to be made with a known probability of being correct.

We should note that rigorous statistical inferences can only be drawn from properly designed experiments. In these, the experimenter has the power to allocate different treatments to particular individuals. Observational studies in which comparisons are made between individuals which happen to have naturally encountered different conditions are always open to qualified interpretation.

11.3 GENERAL PRINCIPLES OF EXPERIMENTAL DESIGN AND EXECUTION

In Chapters 3 and 4 we covered the importance of randomization, replication and blocking in designing an experiment in some detail. We will now see how these topics fit into the broader considerations involved in designing your project.

11.3.1 Why are you carrying out this experiment?

OK – it is a project which fulfils part of the requirement for your degree. However, are you just generally curious about whether changing the nutrients which a plant receives will change its growth rate; do you want to compare the effects of two or more ages of insect host on the rate of parasite development; are you working in the controlled conditions of a glasshouse in the hope that your results may help in the selection of treatments for a subsequent field experiment?

11.3.2 What population are you studying?

You should ensure that the experimental material is representative of the population about which you wish to make inferences. The plant ecologist John Harper has pointed out the 'trilemma' experienced by an experimental scientist. He or she may seek:

1 Precision

Precision can be achieved by using unique genotypes and carrying out the work in controlled environments. This should provide repeatable estimates with narrow confidence intervals but how relevant are the results to the 'real world'?

2 Realism

This may be achieved by studying individual plants in the field. This leads to low precision. Estimates will have very large confidence intervals because of the large amount of random variation. Only large differences between populations will be detected as statistically significant.

3 Generality

It may be desirable to have information about treatment effects on a wide range of, for example, different soil types. With limited resources there is a danger of sacrificing both precision and realism and of ending up with a shallow study.

11.3.3 What are the experimental units and how are they grouped?

Experimental units (test tubes, plots, pots, plants or animals) should not be able to influence each other and should be of a practical size and shape for the study. For example, if you are interested in the weight gain made by insects feeding on different species of plants, the experimental unit is the plant. Sensibly, you may choose to have ten insects feeding on each plant. Their average weight gain will provide a good estimate of the plant's value as a food source but you should not be tempted to consider the weight of each insect as an independent value because they have been competing with each other for food, so you must use the mean weight of the ten insects as the unit of assessment.

The plant pots should be large enough to allow for plant growth throughout the time of the experiment and spaced widely enough apart so that the plants do not shade each other.

It may be possible to group experimental units so that members of the same group will experience similar background conditions. For example, put the biggest insects or the tallest plants into group one, medium sized individuals into a second group and the smallest ones into a third. Consider applying each treatment (and a control) to one individual selected at random from within each of these groups. This will improve precision (in a similar way to stratified random sampling). It will result in narrower confidence intervals and in a greater chance of detecting differences between treatment populations (if they exist).

Treatments should be applied and records should be made group by group rather than treatment by treatment. In the analysis you would then account for variation between these groups or blocks.

11.3.4 What are the experimental treatments?

Is there a naturally defined control? If so, it should be included. For example, in estimating the effects of different levels of a chemical on an organism it is wise to include the standard recommended amount(s) as well as applying none (the control). In addition, you may well apply a wide range of increasing amounts deliberately chosen to include a 'ridiculously' high amount.

Consider carefully how treatments are related to one another. Treatments may appear to be entirely unrelated to each other. For example, you may apply six totally different chemicals. However, if you consider their nature you will often find that there are natural groups within the treatments. Perhaps among six chemical treatments there is a one 'traditional' chemical whose effect should be compared with the average effect of the other five 'new' chemicals.

In contrast, some treatment groupings may be very clear. They may represent increasing amounts or levels of a factor (shade, water or nutrients). It is important to consider the number of levels of the factor, whether it is biologically important for them to be equally spaced or not and how many replicates there should be of each level. In this case the analysis should concentrate on testing whether there is evidence for a steadily increasing (or decreasing) response to increasing amounts of the factor or, perhaps, for a curved response.

Finally your treatments may involve different combinations of factors. As we have seen the efficient way of selecting treatments is to include all possible combinations of all levels of each factor (a factorial experiment, Chapter 7).

11.3.5 Will randomization be across all the experimental units or constrained within groups?

You need to allocate treatments to material at random to avoid bias and to be able to generalize the results. Remember that 'random' is not the same as 'haphazard'. To allocate treatments to experimental units at random requires the use of random numbers. The experimental units are numbered sequentially and the treatments are allocated to them in the order in which the plot number occurs in a sequence of random numbers. A plan of the treatment allocations should be made. This is helpful in making records and may also be used to help explain any unexpected environmental variation which affects the results.

If you decided to group the experimental units into blocks (section 11.3.3) then you should allocate the treatments at random to the experimental units within each block.

11.3.6 What are the aims and objectives of the experiment?

The general aims (purpose) should be clear and the objectives should be lucid and specific and expressed as: questions to be answered (How does pH affect reaction time?); hypotheses to be tested (null hypothesis of no linear response); and effects to be estimated (the mean increase in temperature is 5°C with a 95% confidence interval of between 4° and 6°C).

If you have both a primary and a secondary objective you should make sure that the design of the experiment is effective and efficient for the primary objective and, ideally, also for the secondary objective.

Think carefully about what you will be measuring or counting. Some **variates** will be relatively easy to observe: the number of live seedlings in a pot, for example. Others may be more difficult: the leaf area of a plant, say. Decide which variates are of most interest to you. If they prove to be very time-consuming to record you may choose not to record any others. Consider whether you will analyse each variate separately or whether you will combine any of them before analysis. For example, you might decide to multiply leaf width by leaf length to obtain an approximate estimate of leaf area.

If you make the same measurement on each plant on several occasions (for example measuring height) these will not be independent observations; they are **repeated measures**. A simple approach is to subtract, say, the first height from the last height and to analyse the increase in height over time.

It may well be sensible to obtain some estimate of the size of individuals before they have been exposed to any treatments. For example, the weight of each animal or the height of each plant might be recorded and called a **covariate**. These values can then be used to account for some of what would otherwise be classified as simply random variation present in the experimental results. This may well increase the precision of your experiment. You should ask advice about how to make use of such information.

11.3.7 How many replicates should you have?

This is a common question. To answer it you need to know or be able to make a sensible guess at two things:

1. The minimum size of the difference between any two treatment means which you would regard as being of practical importance (for example, 5 kg $plot^{-1}$ difference in mean yields or 2 mm day^{-1} difference in rate of growth).
2. The likely variability of the experimental material. A pilot study or the results of other people's work on similar systems is useful here. They will probably have presented an estimate of the variability in their experiment in the form of a standard error of the mean or of a difference between two means, a confidence interval, or a Least Significant Difference. All of these contain the standard error of a mean which can be used to calculate the standard deviation. To do this we multiply the standard error by the square root of the number of replicates which they used in their experiment. For example, if the

standard error of their mean was 0.5 kg and there were four replicates then the standard deviation will be: $0.5 \times \sqrt{4} = 1.0$ kg.

The standard deviation can be used to calculate the number of replicates you need to stand a reasonably good chance of detecting the required difference (if it exists) between two treatment populations. You may need to ask a statistician for advice about this because the required calculation depends on the design of the proposed experiment. If you have too few replicates there will be no hope of detecting your required difference. If you have too many replicates you may be wasting valuable time and money. Your project will in all probability have a small budget and you must remain within it.

Once the number of replicates has been fixed it is important to remember to assess and record them separately. Sometimes people inadvertently combine the information about one treatment from all replicates. This makes the results impossible to analyse.

11.3.8 What resources (and constraints) do you have?

Will your project need to take place at a certain time of year? It is no good proposing to study the physiology of hibernating hedgehogs if the field-work must take place in summer! Similarly, many vegetation studies are best carried out in spring and early summer when plants are coming into flower and are relatively easy to identify. If your work is to be carried out in a laboratory it may well be possible to be there only when it is not in use for teaching classes.

Check that you can fit in your project satisfactorily with your other responsibilities, allowing for your taught course timetable and deadlines for continuous assessment work. Do you need transport to get to your experimental site? If so, this needs to be organized and reliable. Does your study need any equipment and, if so, is there any money to pay for it? If equipment needs to be made, is this going to be possible in the time available?

11.3.9 How should you record your results?

Records of results from any scientific study should be accurate and unbiased. Design a record sheet which is clear and easy to use, both when recording and when putting data into the computer (Chapter 4). Each experimental unit should have a unique number. Ideally, this should be used when recording so that you are unaware of the treatment at that stage.

Make sure you know exactly what you are going to record before you start. Make a note of anything odd. Check for gross errors (for example an ant which weighs 200g) and keep the number of significant figures to a minimum.

If you are working out of doors, be prepared for the wind and rain. A clipboard on which the record form is secured by bulldog clips and which is placed in a polythene bag to protect it from drizzle is helpful. A pencil (and a pencil sharpener) is best for writing on damp paper. If you make a mistake, cross out the incorrect value and write the replacement one clearly labelled to one side of it.

You may have access to a tape recorder. This seems an attractive option but they are not infallible and may break down. Also, it may be more difficult to keep track of where you are in your sequence of recording. More recently, hand-held data loggers (small computers) have become popular. Data collected on them can be transferred straight into a computer package like MINITAB via a cable connection between the two machines. They are very valuable if your work involves regular recording of the same experiment. Otherwise the effort required (usually from an expert) to write the program to prompt you for the appropriate data in sequence may be out of proportion to the possible gain in time.

Most importantly, **keep a copy of results in a separate place**. Remember, what can go wrong will go wrong. It is a good idea to use a notebook whose pairs of numbered sheets can be separated by a piece of carbon paper. The duplicate can be torn out and immediately stored in a separate file. Alternatively, use a hardback notebook and make regular photocopies of each page of notes.

Ideally, transfer your results to a computer file immediately after you have recorded them. You should have a minimum of two computer discs with your data on them. One is a **backup copy**. Discs occasionally fail or can be damaged. Professionals have copies of data on three discs. This is to guard against the following possible (if very rare) sequence of events.

You find that the computer cannot read a file from your disc. You assume that your disc has failed and put in your second disc. Unfortunately, the computer cannot read the file from this disc either. It is only now that the possibility dawns on you that perhaps the computer's disc drive is damaging the discs. If you have a third copy on another disc you know that your data are still safe and you will not put this disc into the same machine.

Another reason for having a third disc is that it is good practice to keep one copy in a separate place, in case you lose the two other copies while they are kept together, through fire or theft. If there is a third disc with a friend you can update it, say, once a week as an insurance policy. Also, if you have a brainstorm and remove copies of a file from both of your day to day discs in error (easy to do when you are tired), your third disc is not there so, thankfully, you cannot make the same mistake with that one immediately. If ever you delete a file and regret it because it was your only copy, ask for help from a computer expert immediately. As long as you have not tried to save anything else onto the disc it will usually be possible to retrieve the file.

It is vital to make comprehensive notes of everything that happens as you go along. It is infuriating trying to puzzle out exactly what you did and when some months after the event. Take a few photographs of key features to illustrate the report. For example, striking differences in appearance between the treatments or unexpected events like a flood on part of the site may help to brighten your presentation. Ask a friend to take a photograph of you applying treatments and recording results. When you read a useful paper, make a note not only of its contents but also of its full bibliographical reference (Chapter 12). You may wish to store the references in a file on your word processing package on the computer so that you can edit a copy of this file at the end of your project to provide the references section of your report.

11.3.10 Are your data correct?

It is tempting to assume that you haven't made any mistakes either in recording the data or in transferring them to the computer. This is highly unlikely. It is essential to carry out a systematic check.

First, are all the data present, correctly labelled and in sequence? Are any codes you have used in the correct range? For example, if you have fixed treatments coded 1 to 5 there should not be a 6, and if there are six replicates of each treatment there should not be five codes of '1' and seven codes of '2'. What code have you used for a missing value? (MINITAB uses a *.) Are the missing values correct or should some of them be zeros? Are any of your observations too big or too small? Obvious errors can be spotted by asking for boxplots of each variate when they will appear as outliers. Plots of repeated observations over time can also reveal potential errors. For example, it is highly unlikely that a plant will really have shrunk in height over time.

If you can persuade a friend to help you it is good practice for one of you to call out data from your original notebook and for the other to check it against a printout from the computer file. This will identify any transcription errors. Whenever you have edited your data consider whether you wish to overwrite the previous, incorrect file of data or to save it as a separate file. The number of files you have can grow surprisingly quickly. When you start all seems clear; two months later it is difficult to remember what is in each file. You should plan a consistent system of naming your files. For example, 'e1area05' (there is commonly a limit of 8 letters or numbers for filenames) might stand for experiment 1, leaf area in May. It is important to keep a note of what each file contains and to what extent the data have been checked.

11.3.11 How will you analyse your data?

You should define the specific questions which interest you as particular comparisons before carrying out the experiment. For example does treatment 3 have the same effect as treatment 4? Check that you understand the proposed method of analysis and its constraints and that you will be able to carry it out using whatever you have available, whether this is a calculator, MINITAB on a personal computer, or even perhaps a more advanced package like Genstat, SAS or SPSS. Becoming familiar with your computer package will take longer than you think but will prove very satisfying in the end. If it will take some time before your experiment will produce results, practise the method of analysis using invented data (call your datafile 'rubbish' so that you don't later incorporate the results because you think that they are genuine). This will enable you to gain confidence.

11.4 GENERAL PRINCIPLES OF SURVEY DESIGN AND EXECUTION

Most of the principles we have looked at with respect to experiments also apply to surveys. However there are some points which should be especially stressed as well as some important extra points to consider.

We will illustrate these points by reference to a proposed project involving a survey of the popularity of a collection scheme for different types of waste (glass, paper, batteries, drinks cans and kitchen waste) from households in a small village. The aim is to estimate the level of cooperation in the scheme; to collect views on how it might be improved and to see whether participation in the scheme is related to any characteristics of the households (for example, age of residents and approximate income band).

11.4.1 Is there any bias in sampling?

You may decide to use the voting register to provide a list of all the households in the village. This might be out of date and so exclude the latest housing development from your population. If you decide to sample every tenth household it may be that you accidentally select only ground-floor flats because flats are grouped in two-storey blocks with 5 flats per storey. A random sample is preferable.

If you visit each selected household you may find that in many cases the inhabitants are out when you call. It is important to call back many times (perhaps at a different time of day) to ensure that you catch them in. For example, if you call only between 9 am and 5 pm on a weekday your sample will underrepresent households where all adults are working outside the home. Some people may not wish to take part in the survey. It may be easier, quicker (and safer?) simply to deliver a questionnaire to each selected household. In this case there should also be a covering letter which clearly states where you come from and how you may be contacted as well as the purpose and importance of the survey, together with a stamped addressed envelope for the return of the completed questionnaire. It will undoubtedly be necessary and worthwhile to send duplicate questionnaires and return envelopes to those who have not responded by the date requested. This will minimize '**non-response errors**'; perhaps those who tend not to respond quickly are also those who tend not to participate in the waste collection scheme?

11.4.2 Is your questionnaire well designed?

The design of the questionnaire must be carefully considered. People will be put off if it is very long or if it contains sensitive or ambiguous questions. It is essential to try out your questionnaire on some of your friends first and then on a few villagers to identify any ambiguities from the respondents' point of view.

Ideally, most questions should be answered by selecting an answer from a 1 to 5 scale. For example, 1 = I never do this, to 5 = I always do this, or, 1 = yes and 2 = no. Boxes beside each question on the right-hand side of the page are used for coding the answers (for example, 1 to 5 for an answer and 9 for a missing value or don't know) ready for entering the data into the computer (Fig. 11.1). In addition to such 'closed' questions, you may wish to include a few 'open' questions in which the respondent can say what he or she thinks about the topic. In this case we might ask 'In what ways do you think that the waste collection service might be improved?'

It is important to realize that even if you obtained completed questionnaires from all the households in the village (a **complete enumeration**) there would still be errors in your data. These would not be '**sampling errors**' because we have data from the entire population, but they would be '**response errors**'.

In a postal questionnaire response errors may be caused by:

1. misreading handwriting (age 23 instead of 28);
2. lapses of memory (when did you start participating in the scheme?). It is better to give a range of time periods: less than 6 months ago, between 6 months and 1 year ago, and so on, and ask people to select one;

	Office use only Ref 1–3 ☐ ☐ ☐
1. How many people live in your house? Children (under 18 years old) ☐ Adults ☐ *Please put a number in each box*	☐ 4 ☐ 5
2. Do you put out any waste material for collection by the local recycling group (not the dustmen)? Yes ☐ No ☐ *Please tick one box* *If yes go to question 4* *If no go to question 3*	☐ 6
3. Please say why you do not put out any waste material for collection by the local recycling group: Do not have any suitable material ☐ Too much trouble to sort material and have it ready for collection ☐ *Please tick one box* Any other reason? *Please specify here:* ____________________ ____________________ ____________________	☐ 7 ☐ 8
4. Please place a tick in the box next to each type of waste material which you have placed out for collection in the last month. Newspapers ☐ Magazines/waste paper ☐ Cardboard ☐ Glass ☐ Batteries ☐ Aluminium cans ☐ Tin cans ☐ Kitchen waste ☐ Garden waste ☐ Other *(please specify)* __________	☐ 9 ☐ 10 ☐ 11 ☐ 12 ☐ 13 ☐ 14 ☐ 15 ☐ 16 ☐ 17 ☐ 18

5. What type of house do you live in? *Please tick one box* Terrace house ☐ Semi-detached house ☐ Detached house ☐ Flat ☐ Other *(please specify)* ________	 ☐ 19 ☐ 20 ☐ 21 ☐ 22 ☐ 23
....	
*Thank you very much for completing this questionnaire. We will display a copy of our report in the village shop in May. * Please return your questionnaire in the enclosed stamped addressed envelope.	

Figure 11.1 Questionnaire about recycling household waste in village x.

3. the tendency of people to want to appear 'socially acceptable' and so to overestimate, say, the proportion of household waste they put out for recycling.

A pilot study will help to minimize response errors but it is also important to try and validate the answers independently. For example, you could monitor the actual amount of each type of waste collected from the village each week for a few weeks.

11.4.3 Ethical considerations

Proposed experiments in medical research (**clinical trials**) must be approved by an independent committee of responsible people. For example, the allocation of treatments at random to people who are suffering from a disease may not be considered to be ethical if there is already evidence that one of the treatments is likely to be more efficacious than another. There are also strict controls on research work involving animals. Such matters are the responsibility of your supervisor.

However, there are also more general ethical considerations. It is unethical to waste resources, to carry out a badly designed experiment, or to survey or to analyse it incorrectly and so mislead those reading a report of your results.

11.5 HEALTH AND SAFETY

Your institution will have guidelines which you must read and follow. There will be booklets on this subject available from your library. Half-an-hour spent considering possible problems in advance is time well spent. Accidents do happen but sensible precautions will minimize the risk of their occurrence.

11.5.1 In the field

Field work can be dangerous and you should discuss with your supervisor whether you need to be accompanied or not. You should always leave a note of your route with a responsible person, together with a time by which you should have contacted them to confirm that you have returned safely. You should also always obtain permission to enter and work on private land (this includes Nature Reserves). Plants must not be dug up and removed from a site.

Wear sensible protective clothing and take a map, compass (which you know how to use!), whistle, first-aid kit, and emergency food supplies with you. If your last tetanus inoculation was 10 or more years ago it is sensible to have a booster in case you cut yourself. If you carry out field work near water which is frequented by rats you should be aware of the possibility of contracting leptospirosis. If you work in areas where there are sheep you may risk contracting Lyme disease which is spread by means of ticks. Such diseases may first express themselves by 'flu-like' symptoms, so be aware of these possibilities and contact a medical doctor if you have any suspicions.

11.5.2 In the laboratory

If work in a laboratory is a necessary part of your project you should already have a reasonable amount of experience in this area. You should meet the laboratory supervisor, with your supervisor, and discuss the proposed work. If you plan to use any chemicals which might pose a hazard to health you should complete, in conjunction with your supervisor, a COSHH form to outline the procedures which will be followed. COSHH stands for the Control of Substances Hazardous to Health. The idea is that any possible hazards in your proposed work should be identified. The risk involved is then assessed by considering both the perceived hazard and the probability of its occurrence. For your own protection you should:

Make full use of protective clothing and equipment such as overalls, laboratory coats, safety shoes, dust masks, machine guards.

Make full use of relevant 'control measures' – dust extractors, fume cupboards.

Report to your supervisor any shortcomings in control measures or protective clothing.

Ensure that you understand the job you are about to do and the risks involved.

If you are in any doubt seek advice from your Institution's Safety Officer.

12 Preparing a report. What's it all for?

This chapter provides some guidelines on how to produce a project report. Do check with your supervisor for any particular requirements about presentation and deadlines for your project report. These will usually be listed in a reference document.

There are numerous books and articles which provide advice on how to produce a written report on an investigation (Appendix B). It is valuable to consult several but, since procrastination is one of the most likely problems to beset a writer, you should not get so carried away that the report doesn't appear. The key thing is to put something down on paper. It then provides raw material for revision. A large notebook in which you jot down dates, thoughts, references, methods and results as the study proceeds is invaluable. But remember, the aim is to produce a **report** by the **deadline**.

12.1 COMPUTERS

The utility of using a word processing package cannot be overemphasized. The technique is not just an up-market typewriter on which mistakes can be corrected easily; it allows you to be more creative. It is easier to compose at a keyboard which allows passages to be exchanged, inserted and deleted at will than it is to do so with a pen. If a computer is a relatively new experience for you, do take advantage of the opportunity to use one for your project. When you receive comments on a draft from another student (or in some institutions, from your supervisor) you can edit your report very easily.

Computer packages such as MINITAB enable you to store, edit and analyse your data. You can import summary output from MINITAB into a word processing package like WORDPERFECT. WORDPERFECT can also check your spelling, count your words and suggest alternatives for over-used ones. Other computer packages exist which will check your grammar and suggest how to improve your style.

In recent years computers have revolutionized access to the literature. It is now possible to search databases for published papers which include a keyword or set of keywords. For example, you might ask for references to papers which include the keyword 'blood' or only for those which contain the phrase 'blood pressure'. You can obtain a copy of the output on disc, so that you can read the titles and sometimes the abstracts at leisure and edit out the unwanted references before printing out details of the interesting ones you wish to follow up.

12.2 BASICS

Writers' block is very common. The only way around it is to work back from the final deadline and set down a schedule of work to achieve it. This should include some spare time to allow for emergencies.

12.2.1 Structure

A report should have a clear structure and if you are not sure what that structure should be, start with the basics: useful section headings are:

1. **Introduction** This presents the background to the study, and reviews the relevant published literature before outlining the general aims and specific objectives of the project.
2. **Materials** and **Methods** This contains a description of:

- any pilot study
- the experimental or survey design
- the study populations
- the treatments (including any control)
- the reason for choosing the number of replicates
- the method of randomization
- the methods of accounting for other sources of variation
- the criteria for assessing the outcome of the study
- any possible sources of bias
- any errors in execution (for instance missing observations)
- where and when the study took place
- a description of the apparatus and of any analytical methods used (or a reference to where these are fully described)
- statistical analyses (or a reference to where these are fully described)

You should also state the statistical package which you used (if any). The idea is to enable someone else to be able to repeat your work, if they so desire.

3. **Results** This presents the data obtained with a brief explanation of the major trends revealed. You should illustrate the results section

with tables and graphs (see section 12.3 – illustrating your results) and with reference to statistical analyses as appropriate. Ideally you should provide exact 'p-values' rather than simply '$p < 0.05$'. It is important to concern yourself with your original objectives. It is bad practice to 'dredge the data', in other words to make every possible comparison between treatments in the hope of coming up with a 'significant' difference!

4. **Discussion** How do the results affect our existing knowledge of the subject? The discussion provides the setting for interpretation of your results. You should try to avoid repeating the results in this section. Here you may be more imaginative and speculative in your writing, perhaps outlining possible reasons for the findings and giving suggestions for further work. It is important to check that your conclusions are justified by the analysis of the data and that you are not trying to make the results fit some preconceived idea of what you think ought to have happened.

   This section is the place also to compare and contrast your results with those of previous studies, perhaps suggesting reasons for any discrepancies. It is important not to attempt to extrapolate the results of a sub-population which you have studied to the whole population. For example, what you have discovered about the behaviour of worms on the local farm may well not be true for worms of a different species, or on a different soil type, or 200 miles further north.

5. **References** This is a list of the material cited in the report. You should quote sources of reference from the published literature or personal communication in the text where appropriate and list them fully in the References section. You may be instructed to follow a particular 'house style'. This may seem fiddly to you but it is essential to give accurate references and to double check that the list contains all the required references and no others. Consider how annoyed you would be if you wanted to follow up an incorrect reference.

6. **Acknowledgements** You should thank all those people or organizations who have helped you: your supervisor, members of the statistical and computing staff, the laboratory supervisor and your friends. Don't forget the landowners if you have been working in the field and do send them a copy of your finished report.

7. **Appendixes** These are optional and contain items which are not required to follow the flow of the argument. Examples include: raw data, calibration curves, species lists and mathematical equations.

8. **Abstract** A brief statement of aims, methods and results is helpful in all but the briefest of reports. It should focus on the 'take-home message'. This may well be the last part which you write. In many journals this is placed at the beginning of the paper so that it can be readily scanned by those who may not have time to read any more.

12.2.2 Drafts

When the above outline is in place, a first draft can appear. At this point it becomes obvious that there are gaps which you need to fill or that points should be made in a different order to emphasize connections. It is important to immerse yourself in this task so that the momentum is maintained. Write 'REPORT WRITING' in your diary and put a notice saying 'REPORT WRITING – KEEP OUT' on your door! A first draft will say things like '**must find comparative figures** or **check analysis** or **find reference** or **yuck! – rewrite this!**'. These points are attended to in preparing a second draft which should then be given to a friend whose opinion you value and (if it is the custom) to your supervisor for comments.

This is a vital stage and such comments will lead to great improvements in the clarity of the third draft. Don't be surprised if the reviewer covers it with comments in red ink. We all have to endure this because it is easier for others to spot our errors and inconsistencies than it is for us to see them ourselves. You can usefully have your revenge by constructively criticizing another student's draft report.

As you are writing you should consider who will read your report. What will they know already? What do you want them to understand and to remember? Browse through the literature in your field. When you find a paper which is clear, brief and stimulating, use it as a model for your approach.

When you find yourself absorbed in typing your report onto a computer you should be sure to take a short break after an hour's work, and a long break after a further hour, otherwise you will find that your body will protest, your mind become tired, and you will make mistakes. If you find that your neck is beginning to ache ask advice about your body position relative to the computer. Perhaps the chair needs adjustment.

12.3 ILLUSTRATING RESULTS

The question of whether to use graphs or tables can be a difficult one to resolve.

12.3.1 Graphs

Graphs are excellent for communicating qualitative aspects of data like shapes or orders of magnitude. They are invaluable in exploring the data. For example, they can help you to decide whether a straight line model is sensible or whether there are outliers. Graphs are also helpful in showing a broad picture, or in reminding a reader of principles.

It is important to label the axes of the graph, giving the units of measurement, and to present a suitable legend (caption). When you present a graph you should also include a measure of the variability of the data. This is commonly achieved by presenting a vertical bar labelled 'SE mean' (standard error of a treatment mean after analysis of variance, Chapter 7). Remember that when each treatment has an equal number of replicates the same SE mean applies to each treatment and so only one SE bar is needed on the graph (Fig. 12.1). Alternatively, you may put a least significant difference (LSD) on your graph or you may put 95% confidence intervals around each mean. It is important to make clear which of these three possible measures of variability you have used.

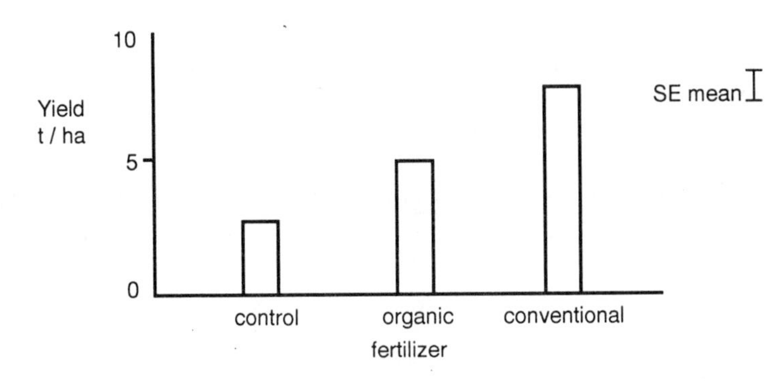

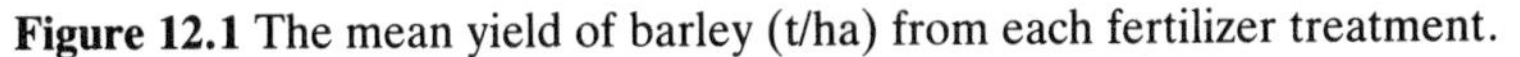
Figure 12.1 The mean yield of barley (t/ha) from each fertilizer treatment.

12.3.2 Tables

Although graphical plots can show general trends, tables are required to convey quantitative features. Ideally, a good table will display patterns and exceptions at a glance but it will usually be necessary to comment in the text on the important points it makes. Some useful guidelines for producing tables have been suggested by Ehrenberg*:

1. Round data to two significant figures. This refers to digits which are 'effective', in other words, which vary from one observation to another.
2. Provide row and column averages or totals on the right-hand side of the rows and at the bottom of the columns.
3. If you want to compare a series of numbers it is easier to do so if they

* Ehrenberg, A.S.C. (1977) Rudiments of numeracy. *Journal of the Royal Statistical Society*, A, **140**, 277–297.

are in a column rather than in a row. This is because the eye finds it easier to make comparisons vertically than horizontally.

4. Ordering columns and rows by the size of the values in each cell helps to show any patterns.
5. Single spacing between cells guides the eye down a column and gaps between rows guide the eye across the table.

If you have carried out a factorial analysis of variance there is a neat way of presenting the treatment means in a table, with their standard errors. You may remember that in Chapter 7 we discussed a wholly randomized design experiment with two levels of sowing and two levels of cutting the field margins. The factorial analysis of variance of the mean number of spiders per quadrat in each plot produced an error or residual mean square of 1.333 (p. 62). Each of the four treatments had four replicates, so each main effect has eight replicates. Use this information and the formula for calculating a standard error (above) to check the following table:

Table 12.1 The effect of sowing and cutting on the mean number of spiders per plot

	sown	unsown	mean	SE cutting mean
cut once	19.5	16.5	18.0	
cut twice	15.0	13.0	14.0	0.408
mean	17.25	14.75	16.0	
SE sowing mean		0.408		
SE treatment mean		0.577		

12.4 LANGUAGE

The report must be legible and well-written. It is normal in scientific reports to write in the past tense rather than in the present tense and to use 'we' rather than 'I'. It is traditional to use the passive rather than the active voice although many journals are now starting to encourage the latter. So you should write 'the vegetation was sampled (or we sampled the vegetation) using five randomly positioned quadrats in each plot', rather than: 'I will choose five quadrats selected at random from my field plots'.

Generic names of organisms take an upper-case initial and specific and subspecific names commonly have a lower-case initial and both are always in *italics* (or underlined): *Beta maritima* ssp. *maritima*. To be complete you should follow the binomial name by its authority (an abbreviation for the name of the person credited with describing the species first): *Lolium perenne* L. (L. stands for Linnaeus). You should check spelling with a standard list and state which list you have used in the materials and methods section.

12.5 EXAMPLE OF A SHORT REPORT

Read the following example of a short report and comment on its strengths and weaknesses.

The effect of fertilizer on forage rape: a pilot study
A.N. Other, Small College, Largetown, Midshire

12.5.1 Abstract

An experiment was carried out to investigate the response of forage rape seedlings (*Brassica napus* L. ssp. *oleifera*) to the major nutrients on a soil which was thought to be nutrient-poor. It was concluded that, for this soil, the addition of 67.5 g/m^2 of a N:P:K:7:7:7 compound fertilizer would be adequate and that the effect of lower levels should be investigated in a subsequent pot experiment before carrying out a field experiment.

12.5.2 Introduction

Forage rape (*Brassica napus* ssp. *oleifera*) is a crop grown for feeding to livestock in autumn and winter. An average fertilizer requirement for the crop is 125 kg ha^{-1} N, 125 kg ha^{-1} P_2O and 125 kg ha^{-1} $K^K{}_2O$ (Lockhart and Wiseman, 1978). We wished to investigate the optimal amount of a compound fertilizer containing these nutrients to apply to seedlings growing on what was reputed to be a very nutrient-poor soil. A pilot study was carried out to find out the range of amounts of fertilizer which should be used in a more detailed pot experiment before carrying out a field experiment.

12.5.3 Materials and methods

On 10 September 1991 a 1 cm layer of gravel was put into each of 80 pots (10 cm × 10 cm in cross section and 12 cm deep). The pots were then filled with sieved soil from a nutrient-poor site near Largetown, Midshire. There were 20 replicate pots for each of four fertilizer treatments: the equivalent of 67.5, 101, 135 and 270 g m^{-2} (1 g m^{-2} is equivalent to 10 kg ha^{-1}) of Growmore granular fertilizer (N:P:K::7:7:7) which was ground to a powder and mixed throughout the soil. The recommended rate for an 'ordinary garden soil' is 135 g m^{-2}. The treatments were identified by colour codes (unknown to the assessors to prevent bias) painted on the pots. The pots were arranged in a wholly randomized design in the glasshouse.

Ten seeds of *Brassica napus* were sown in each pot and covered with a thin layer of soil. The pots were examined daily and watered using rainwater as required. The plants were thinned to six per pot on 16th September and to 4 per pot on 20th September.

At harvest on 15th October the number of plants per pot was counted and their health scored (0 = perfect to 5 = dead). Plant height and the number of true leaves per plant were noted and leaf width and length were put into a formula to obtain an estimate of leaf area (H. Moorby, pers. comm.)

$$LA = -0.35 + 0.888\,(LW \times LL)$$

where LA = leaf area, LW = maximum leaf width and LL = midrib length.

Shoot material from each pot was put into a labelled paper bag and oven-dried at 80°C for 48 hours before being weighed.

This report concentrates on the effect of fertilizer on shoot dry-matter yield. The data were analysed using analysis of variance in MINITAB Release 8 (MINITAB Inc., 1991).

12.5.4 Results

By 16th September germination had occurred in all pots but with increasing fertilizer there were fewer plants and these were of smaller size and were yellower. There was very strong evidence that fertilizer affected shoot dry matter yield ($p < 0.0001$, Table 12.2).

Table 12.2 Analysis of variance for total shoot dry-weight per pot

Source	DF	SS	MS	F	P
Fertilizer	3	1.10257	0.36752	29.61	0.000
Error	76	0.94326	0.01241		
Total	79	2.04584			

Shoot dry weight decreased with increasing fertilizer (Table 12.3, Fig. 12.2).

Table 12.3 Mean shoot dry-weight yields per pot

Fertilizer	gm^{-2}	67.5	101	135	270	SE mean
Dry weight	g	0.4770	0.3655	0.3235	0.1505	0.02491

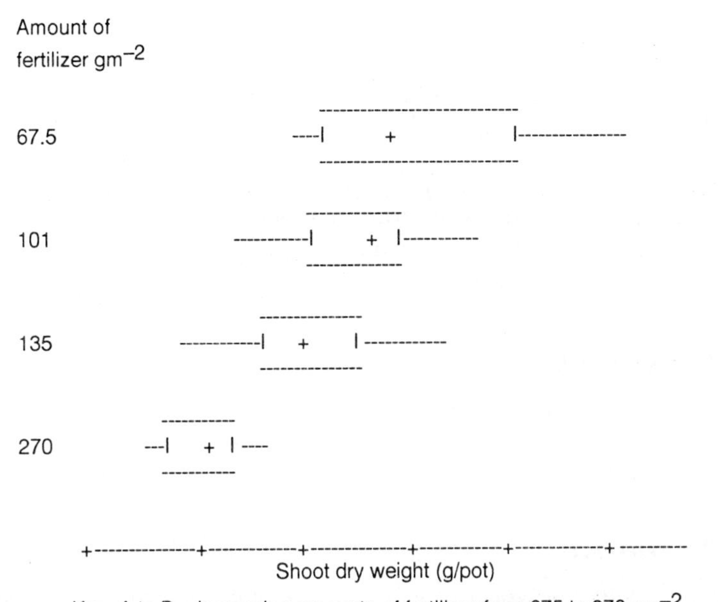

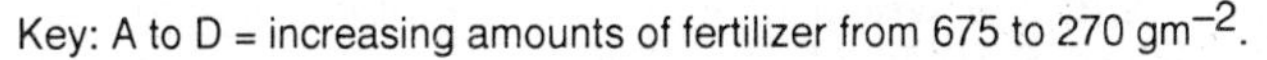

Key: A to D = increasing amounts of fertilizer from 675 to 270 gm−2.

Figure 12.2 Boxplots of the total shoot dry weight per pot for the four fertilizer levels. There was a positive linear relationship between increasing leaf area and increasing shoot dry weight (Fig. 12.3).

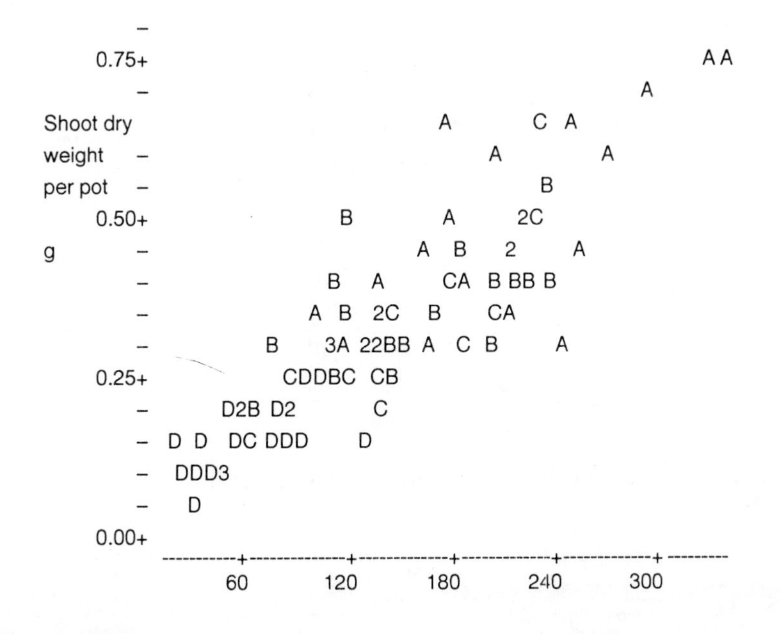

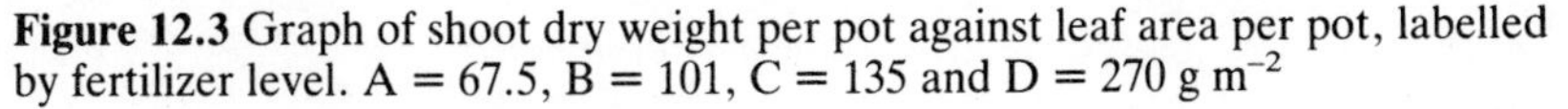

Figure 12.3 Graph of shoot dry weight per pot against leaf area per pot, labelled by fertilizer level. A = 67.5, B = 101, C = 135 and D = 270 g m^{-2}

12.5.5 Discussion

The soil tested was of a surprisingly high nutrient status as indicated by the performance of the plants in the pots receiving the lowest amount of fertilizer, which gave the highest yield. Greater amounts had a toxic effect. Perhaps even less than 67.5 g m^{-2} would give an even higher yield. For the next pot experiment, levels of fertilizer should cover the range 0 to 120 g m^{-2}. It is possible that plants grown for longer than one month might have a greater fertilizer requirement.

12.5.6 Acknowledgements

I thank the glasshouse staff for maintenance of the plants and the first-year students on Course 199 for harvesting them and recording the data, with help from P.Q. Smith. I thank E.D. Jones for computing advice.

12.5.7 References

Lockhart, J.A.R. and Wiseman, A.J.L. (1978) *Introduction to Crop Husbandry*, 4th ed, Oxford, Pergamon Press, pp 302.

MINITAB Inc. (1991) MINITAB Reference Manual. Release 8. PC version. MINITAB statistical software, PA USA.

Appendix A Choosing how to analyse data from a replicated, randomized experiment

First plot the data – histograms, stem-and-leaf diagrams, boxplots, plots of one variable against another. Then choose from the bulleted options below.

- **Data are continuous** (kg, mm, etc. or, if they are counts, these are large values per experimental unit).
Use analysis of variance (with blocks, if appropriate) to compare treatments:

i) Consider whether there is a factorial structure – if so account for main effects and any interactions in your model.
ii) Consider whether a factor is present at increasing levels – if so use linear regression to see whether there is a linear response or not.
iii) Test the assumptions of analysis of variance/regression by plotting histograms of residuals (to check normality) and residuals against treatments (to check for similar variability).
iv) If the data do not meet the assumptions, either ask advice about transforming your data or consider a non-parametric test (see next).

- **Data are scores or ranks or are continuous but from skewed populations.**
Use non-parametric equivalents of analysis of variance (Mann-Whitney for two treatments, Kruskal-Wallis or Friedman for more than two treatments, the latter with blocks).

- **Data are categorical (each observation can be placed in a category)**
Use chi-squared contingency test to compare the proportion of individuals in a category for the different treatments.

Appendix B
Further reading

There are vast numbers of books written about statistics and carrying out research at a wide range of levels. I have selected a few which are modern and relatively easy to understand.

ELEMENTARY STATISTICS

Campbell, R.C. (1989) *Statistics for Biologists*, 3rd edn, Cambridge, Cambridge University Press.
A revision of a classic textbook which covers all the topics in this book except PCA but at a more technical level, including some MINITAB instructions (and ones for two other statistical packages, Genstat and SPSS).

Clegg, F. (1982) *Simple Statistics. A course book for the social sciences*, Cambridge, Cambridge University Press.
An excellent, simple and humorous book.

Mead, R., Curnow, R.N. and Hasted, A.M. (1992) *Statistical Methods in Agriculture and Experimental Biology*, 2nd edn, London, Chapman and Hall.
A revision of a very popular textbook. It is more sophisticated and covers more complex methods than the present book but is highly recommended as a next step if you have found this book easy to follow.

Neave, H.R. and Worthington, P.L. (1988) *Distribution-free Tests*, London, Unwin Hyman.
This covers non-parametric statistical methods which are useful when your data do not meet the assumptions (like normality of residuals) required for parametric tests.

Porkess, R. (1988) *Dictionary of Statistics*, London, Collins.
Provides definitions of technical terms.

Rees, D.G. (1989) *Essential statistics*, 2nd edn, London, Chapman and Hall.

Very clear. Covers many of the same topics as this book but with a slightly more mathematical approach; omits analysis of variance and PCA but includes a final chapter on MINITAB examples.

Samuels, M.L. (1989) *Statistics for the Life Sciences*, San Francisco, Maxwell Macmillan.
Very thorough. Includes basic analysis of variance and linear regression but not PCA.

Sokal, R.R. and Rohlf, F.J. (1981) *Biometry. The Principles and Practice of Statistics in Biological Research*, New York, W.H. Freeman and Co.
More advanced; many research scientists would regard this as their standard reference book and it is fine if you are really keen.

MEDICAL STATISTICS

Any one of these three texts would provide you with a straightforward explanation of most of the subjects covered in this book (and a few others), but in the context of medical research.

Campbell, M.J. and Machin, D. (1990) *Medical Statistics – A Commonsense Approach*, Chichester, John Wiley and Sons.

Mould, R.F. (1989) *Introductory Medical Statistics*, 2nd edn, Bristol, Adam Hilger.

Petrie, A. (1987) *Lecture Notes on Medical Statistics*, 2nd edn, Oxford, Blackwell Scientific Publications.

EXPLORING AND INTERPRETING DATA

Afifi, A.A. and Clark, V. (1990) *Computer-aided Multivariate Analysis*, 2nd edn, New York, Van Nostrand Reinhold.
More advanced; includes principal components analysis.

Anderson, A.J.B. (1989) *Interpreting Data*, London, Chapman and Hall.
A useful next step; more advanced.

Chatfield, C. (1988) *Problem Solving – a Statistician's Guide*, London, Chapman and Hall.
Very useful for developing a 'feel' for the problems involved in analysing real-life data.

Marsh, C. (1988) *Exploring Data. An Introduction to Data Analysis for Social Scientists*, Cambridge, Polity Press.
Excellent for complete beginners with an arts or social science back-

ground, but concentrates on descriptive statistics rather than testing hypotheses.

MINITAB

MINITAB Reference Manual PC version. Release 8. (1991) PA USA. MINITAB. Inc.
Useful to browse through and see what MINITAB can do. (Remember that you can obtain most of this information by using the HELP command (F1) on the computer.)

Monk, A. (1991) *Exploring Statistics with MINITAB. A Workbook for the Behavioural Sciences*, Chichester, John Wiley.
A useful adjunct to the MINITAB manual with examples of the methods used in this book (except PCA and analysis of variance).

Rees, D.G. (1989) *Essential Statistics*, 2nd edn, London, Chapman and Hall.
Final chapter contains numerous examples of MINITAB analyses.

PROJECTS

Bell, J. (1987) *Doing Your Research Project. A guide for first-time researchers in education and social science*, Milton Keynes, Open University Press.
Excellent practical advice – albeit with a social science bias. Includes principles of questionnaire design.

Chalmers, N. and Parker, P. (1986) *The OU Project Guide*, Taunton, Field Studies Council.
Well worth consulting if you are carrying out an ecological project in the field.

Fowler, J. and Cohen, L. (1990) *Practical Statistics for Field Biology*, Milton Keynes, Open University Press.
Very popular, especially with ecologists; a slightly more mathematical approach than this book.

Pentz, M., Shott, M. and Aprahamian, F. (1988) *Handling Experimental Data*, Milton Keynes, Open University Press.
A beginners' guide: very useful.

REPORT WRITING

Cooper, B.M. (1964) *Writing Technical Reports*, Harmondsworth, Penguin.

Particularly helpful on correctness and style.

O'Connor, M. (1991) *Writing Successfully in Science*, London, Harper Collins Academic.

An excellent book which covers all aspects of communicating in science, including presenting posters and talks as well as writing a paper.

Wheatley, D. (1988) *Report Writing*, London, Penguin.

A good starting point.

OTHER BOOKS

Do look at any reference books which your lecturers recommend. They will, no doubt, have good reason for making their suggestions. It is very helpful to read about the same method of approaching a problem in several books as different authors have different approaches and it is important to find one which makes the topic clear for you.

Index